EXTINCTION

EXTINCTION
Evolution and the End of Man

Michael Boulter

Columbia University Press

New York

In memory of my parents
Harold Boulter, 1898–1961
Dorothy Boulter, 1904–1973

COLUMBIA UNIVERSITY PRESS
PUBLISHERS SINCE 1893
NEW YORK
Copyright © 2002 by Michael Boulter

Library of Congress Cataloging-in-Publication Data

Boulter, Michael Charles.
Extinction : evolution and the end of man / Michael Boulter.
p. cm.
Includes bibliographical references (p.   ).
ISBN 0-231-12836-3 (cloth : alk. paper)
1. Extinction (Biology) 2. Evolution (Biology)
3. Nature—Effect of human beings on. I. Title.
QE721.2.E97 B68 2002
576.8'4—dc21   2002034783

Columbia University Press books are printed on permanent
and durable acid-free paper.
Printed in the United States of America.

C 10 9 8 7 6 5 4 3 2 1

Contents

	Introduction	vii
1	The Past Is Not Over	1
2	Extinction	23
3	A System out of Chaos	56
4	From Dinosaurs to Us	88
5	What's in a Name?	125
6	A Man-Made Extinction Event	157
7	Humans and the Future	177
	Notes	194
	Acknowledgements	203
	Index	205

Introduction

An e-mail message showed up on the screen when I logged on after breakfast. For my taste, boiled fish and rice with chopsticks don't make the best start to the day, but the message more than made up for that. It was manna from heaven, like scrambled egg and smoked salmon, a taste of something good in this unfamiliar country.

The message was from one of my students in London, Dilshat Hewzulla. He's a computer scientist from north-west China, a whiz kid at taking a complex system from one domain and computing it into something recognisable in another. I give him huge databases from extinct and living plants and animals, as well as from the environment, and he finds mathematically significant patterns for evolutionary and environmental interpretations. It may sound simple but it isn't. In his e-mail Dilshat explained that in my absence we had received a letter from The Royal Society publications office saying that our manuscript had been accepted by the editor. Apparently two of the referees had waxed lyrical about it and the other had said it was rubbish and shouldn't see the light of day. But our arguments had won through. My research group had finally gained the recognition of others and was seen to be on to something that would interest the rest of the scientific community.

With an excited feeling of satisfaction, I went off for a cup of tea. That, at least, they do have in Taiwan. I was in the capital, Taipei, at the Academy of Sciences, as the UK delegate to the General Assembly of the International Union of Biological Sciences. The tearoom was not crowded. Across the room my friend from the United States waved me to join his little group of three. They were talking about the major topic of the meeting: the effect of man changing the environment.

The day before, the Assembly had charged two of the scientists seated around that table with the task of finding ways of monitoring these changes upon the world around us. There had been hints from some delegates that without a clear programme of measurements their governments wouldn't take the United Nations Biodiversity Convention seriously. 'How can a forum like this agree on what to measure?' asked one. 'The diversity of life and environment is so big and complex, it's hard to know where to start, and most species are not even described.' 'How can we bring marine and terrestrial ecologists together, whose different domains influence one another very strongly?' 'Do meteorologists ever speak to biologists or those modelling sea-level changes?' Not only are there these scientific problems but also the political ones, as well as financial issues. 'The developing world will never cooperate as long as the West carries on using so much fossil fuel,' said the Indian delegate, 'and even you developed nations can't raise the money to educate people about biodiversity, let alone monitor the changes in environments.'

Later, the Assembly did agree on some priority locations. They designated areas around the world where individual animals and plants are to be identified and counted, the rainfall and temperature logged and the soil types classified. But there are so many other variables to monitoring these possible changes, and they are all affected in different ways at different rates. The scale of time to be monitored was another factor to include in the survey: changes overnight; throughout the year; a decade; a millennium; a few million years. Though my two friends are leading specialists in mainstream biology and ecology, on their own they didn't have a clue how to begin to advise their Ministries. Theirs was an intractable dilemma. We know so little about the losses of species and ecosystems over the last hundred years, yet we all have a strong gut feeling that we are wrecking life on our planet. For governments, neither the questions nor the answers will win many votes.

But they are questions we must not shirk just because of their intractable complexity. They are not for experts in any one discipline, such as my friends in the tearoom, but for the person who can connect all the many disciplines involved. My attempt to contribute some kind of answer to this dilemma is the narrative of this book, and Dilshat's e-mail message signalled the first attempt at a new approach. Our Royal Society paper brought together data from biology, ecology and geology and analysed them mathematically. The data come from changes that have been taking place through tens, hundreds, thousands and millions of years, and the patterns of biological diversity that emerge through evolutionary changes help understand many of the problems raised in the Taipei tearoom. The patterns our study revealed are journeys through time. The journey includes extinctions, origins and diversifications of species and larger groups. The biology matches changes that can be seen in environments, drawing the two different subjects closer together. These connections not only have an exquisite beauty when enshrined on Keats's Grecian urn but also they are seen to have changed through the scales of geological time: ' "Beauty is truth, truth beauty" / – that is all ye know on earth, and all ye need to know.'

This is the story of evolution on our planet from the time of one set of extinctions to another, which covers the last 65 million years and reaches into the future. As well as clues from fossils the plot has evidence from every witness we can find – modern plants and animals, rocks, chemicals, atoms and other sources. The information and ideas come from most of the natural and physical sciences and beyond. We bring together as many of these natural processes of environment and biology as we can, as well as evidence from many hitherto separate disciplines.

I used to call myself a paleontologist because I studied fossils. My research thesis work in the 1960s looked into thousands of broken-up bits of trees that turned out to be 6 million years old. They came from a clay pit in the middle of England, a collapsed cave in the limestone hills of Derbyshire, and they showed that a warm redwood forest had once grown there. It was not unlike parts of California today but with an English scale to the landscape. So much has changed in such a geologically short period of time, yet the Peak District really did look like *Sequoia* National Park in California, the dense mixed forest of deciduous English oak and elm mixed with trees now native only in North America, as well as others found mainly in the East, making rich soils and colourful hills. Where there was bad soil and good drainage, heathland was much like that today; where there was wetland different but familiar plants and animals prospered.

This was a few million years before the glaciers of the ice ages penetrated into the presently highly populated temperate regions of the Earth where most of the large cities of the developed world are now situated. Before the glaciers spread from the poles, warm temperate forests dominated humid hilly landscapes. Shrubs and heath covered the drier soils and grass became widespread for the first time, encouraged by vast increases in the numbers of stooping grazing mammals. The atmosphere had more CO_2 than now, which meant it was much warmer. The polar icecaps were much smaller than now, but growing. If you had taken an aeroplane journey with the usual flight paths from Europe to California, or over Siberia to Beijing, you would have seen the same kinds of forest on both trips: redwoods and warm temperate forest on the hills, swamp cypresses and shrubs near the water, like the kind you see now in the Florida Everglades. They were mixed in with more of the kinds of trees and shrubs you see in the southern United States and warm temperate China – oaks, maples, pines. Further north, a flight from Heathrow to Vancouver would cross the cooler landscape of familiar birch forest with pine and alder: there was much less sea than now, and no ice.

Over the last forty years specialists from different parts of the natural sciences have come together to paint pictures of how the world looked millions of years ago. Their reconstructions are opening up new issues about climate change, plant migration, evolution, ecology and the statistics of populations. Now that we are taking these previously separate disciplines together, we can begin to see how they affect the urgent new environmental issues facing our modern world. My own studies have strayed into many other different methodologies, genetics, geology, ecology, taxonomy and even statistics. So now I call myself an evolutionary biologist rather than the paleontologist that I once was.

Since those exploratory days of the 1960s the scientific literature has been filled with detail about these biological and environmental changes. As the first half of the twentieth century was for theoretical and descriptive biology, so the second half accumulated large amounts of data about the evolutionary relationships between species and the environment. A climax was reached in 1977 when Fred Sanger learnt how to sequence the gene leading to genetic engineering. Two decades later DNA sequencing is being done automatically. Now the results create huge new databases each day, like new rows of books on shelves a kilometre long. Environmental ecologists and taxonomists are part of this new age of very large data recovery, and are beginning to seek automatic techniques to find, store and analyse the data. But there's so much becoming available that it's hard to manage at an international level.

One of the achievements of my research group at the University of East London is that we can analyse large databases interactively across the internet. This means we are able to assemble ideas and information from different sources, from which we can begin to see life on our planet as a complex system. Maybe we can understand how it survives and changes as a whole, for only then can the problems raised around the Taipei tea table be understood and tackled. We are just beginning to make a start to link the bits together.

Throughout my career I have been privileged to see many more of these bits from the whole of science than most people. At University College London during the early 1960s my degree course included lectures by Francis Crick and Sydney Brenner just when they were cracking the triplet genetic code. This was also when J. B. S. Haldane, the pioneer of an earlier revolution in genetics, was to be seen walking through the North Cloisters with a pillow stuffed up his jumper to comfort his cancer. The mathematics of mutation and the recombinations in dominant and recessive gene characters were the centre of his kind of biology: it is hardly heard of today. From this cusp between the old and the new I wandered with chemistry and geology, got stuck in thermodynamics, and touched on the philosophy of science. The more adventurous natural scientists were looking outwards from the strict demarcations of individual disciplines. J. Z. Young was busy changing the way anatomists think and Peter Medawar was breaking thoughts about immunology and the way science works. Karl Popper was down the road and the Beatles slept on the floor next door in my student residence.

All these different perceptions of life were being assembled together in the same place at the same time, squeezing out traditions from separate backgrounds into one amorphous shape. At least it seemed to be amorphous then, hard to put into any clear context or application. They were years of joy for the fearless intellectuals of science. Now, just forty years on, the shape has a much clearer identity, itself being replaced by a fresh wave of integration with new objectivity. Studies of organismal biology had peaked by the 1960s. The principles of structure and function which gave names to 'genes' suddenly led into the language of the triplet code in molecular genetics.

I was helped by my tutor, Bill Chaloner, a luminary in paleontology with a gift for communicating the fascinations of evolutionary processes. Those critical studies led into more defined ideas on how landscapes and ecologies change through different timescales. We were fired by the enthusiasms of the new wave to link all these traditions by looking at an issue from several different perspectives. They were exciting times because you could feel attitudes changing. That's happening again, now, at the beginning of the new century. But this time the changes are going to be very big indeed and are beginning to affect our lives.

These experiences have influenced me to give a broad mind to an argument, often at the risk of being called 'ecumenical'. In retrospect I see that's how I reacted to the many factors relevant to environmental and evolutionary biology, genetics, geology, ecology and mathematics. They are all working together, constituting a complex system on our planet that can be traced back to the extinction of the dinosaurs and beyond. For most of the incidents that are thought to have changed that system there are a number of opposing theories. For example, to explain the sudden demise of so many large groups of animals 65 million years ago, there are at least four different ideas. First, a meteorite hit the Earth causing a 20km crater just off the Mexican coast and world-wide fires that killed big animals. Second, there was severe volcanic activity in India. Third, there were continuing physiological difficulties controlling body temperature. Finally, all the food ran out and the dinosaurs starved to death.

The different theories are a good example of how science works, with argument and limited facts to test the more fanciful solutions being offered. What holds good as an answer today is more than likely to be different from what was understood yesterday, and it will differ again tomorrow. But trends and patterns do emerge, and we are beginning to see things more clearly with more data from different disciplines. The growth of computing power and the introduction of the internet have been vital factors in making these leaps in understanding possible.

On the other hand, these are frightening times. On New Year's Day 2000 the World Wildlife Fund and the *Guardian* newspaper published a booklet entitled *A New Century a New Resolution*. In it, the then chief scientific adviser to the UK government, Sir Robert May, issued a warning about 'our greatest challenge'. In his view, this was to 'ensure that any increase in global productivity is achieved in a sustainable and environmentally friendly way. We really do live at a special time in the history of Life on Earth. A time when human activities have come to rival the scale and scope of the natural processes which built, and which maintain, the biosphere.' If we take action now, he argued, we can avert a catastrophe.

I take a more pessimistic view. In this book I present evidence to show that that catastrophe has been well under way for many thousands of years, and that Bob May's observations of what we are doing to the environment now amount to just a final nail in the coffin. Most people think of time only in terms of their own lifelong experience of it, or a few hundred years more, at best. But if you extend your thoughts of time back further, past the last millennium into the first, and think of what the world might have been like, things look quite different from today. Now compare our present world with what it looked like before humans started to interfere, a few thousand years ago, and there are more changes still. The evidence of what early human hunters did to other large mammals shows enormous horror, resulting in the extinction of a number of species. It has all happened since the end of the last full glaciation, 10,000 years ago. That's a period of time few of us are ever asked to consider. So when I'm asked to say how long it will be before the forthcoming extinctions, I say: 'Soon, but remember, I'm a paleontologist.'

I

The Past Is Not Over

Some remnants of Iron Age man

Most people have some kind of overview of the history of the last thousand years, but few can see back much further. Tirefour Broch, an Iron-Age fortification, stands on the island of Lismore in western Scotland. Archeologists claim it is about 2,100 years old, a 12-metres-diameter stone circle still standing 3 metres high (see figure 1.1). The broch is on the highest part of the long island and has spectacular views over the waters of Loch Linnhe to Fort William and Ben Nevis in the north and the Isle of Mull in the west. But these views are very different from those seen by the fort's builders and inhabitants, protecting themselves from brigands attacking from the mainland forests, their sheep and chickens safe from the mainland's wolves. Then, the mountains and the lowlands down to the sea were covered with dark forests of Scots pine, oak, birch and alder. Above the tree line grew the now widespread purple heather, holly and cranberries, mixed in with small shrubs. The animals that roamed this area were diverse and fierce. It was not a good place for the new humans migrating there from the south and east for the first time. At least the winters were warmer than on mainland Europe. This was due in part to the oceanic climate from the Atlantic Ocean's Gulf Stream that wound up the west coast of England towards Lismore.

Scottish brochs such as Tirefour were built a few hundred years later than the pharaohs' pyramids in Egypt when the young Tutankhamun was king, a civilisation much more advanced than Scotland's. The contrast

Fig. 1.1 Tirefour Broch, Isle of Lismore, Argyll, Scotland. The broch is about 12 metres in diameter and 3 metres high at its tallest point. To the left is an entrance to the inner chamber. The first century BC is the most likely age, based on C¹⁴ dating, but 500 and 600BC have been suggested from earlier datings.

between this thinly populated island and the sophisticated communities of Egypt along the banks of the Nile is marked. In Scotland it was a hard, cold life with continual threats from other humans reaching the end of their migration trails and from hungry animal species. In northern Africa, the warm open grassland prairie and desert scrub teemed with mammals such as gazelle, ass, hartebeest and hyena.

The sight of these prairies was not far removed from what you still see in nature reserves further south on that continent in Kenya down to South Africa. We know the mammals that roamed further north from paintings found in Tutankhamun's tomb showing the king hunting in his chariot with courtiers, fan-bearers and bodyguard. It was easy to settle in the river valleys where whole communities had become established and the treasures that survive along the river Nile bear witness to their cultures.

At the same time, far northwards the cold winters made life a lot less comfortable for the migrating humans. By then mammoths and other large mammals had become extinct, killed off by human hunters, after they had taken millions of years adapting to the more extreme conditions. The new humans were trying to acclimatise much more quickly, helped by their intelligence and skills to make spears, daggers, clothes and shelter. The broch on Lismore, and twenty or so others like it on neighbouring islands, is among the oldest buildings in Europe. Although most of them are being broken up by the weather and more recently by vandalism, they are being preserved by slow immersion into the landscape of pasture. Enough evidence survives to allow us to reconstruct much of the way of life of the Iron Age settlers, their hunting, culture, religion, their relationship with the natural environment. Life was hard but becoming ordered, barracked within these communes. Archeologists are also discovering bronze swords and other weapons at contemporary sites along the Atlantic seaboard of western Europe, which suggests that trade and fighting were part of their life too.

Entry to the broch was by a very narrow passage to a central courtyard where meals were prepared, livestock fed, and groups worked together. Around this area, next to the inside of the wall, was a two-storey wooden structure roofed with straw. This was split into small chambers for sleeping and storage. Variations of this kind of structure were built by other early human groups in different continents as man and civilisations migrated. The communal kasbahs in Morocco are still inhabited, with the central courtyard for animals, and the occupants cooking, sleeping and defending in rooms around the sides. In Kazakhstan partitioned tents serve the same purposes for still migrating people. As well as being good defensive fortresses against other human groups and tribes, they also offered protection from animals of prey threatening their livestock.

Men farmed barley and hunted together in teams, women looked after livestock, cooked, and nursed children. There was little privacy for cosy family gatherings. Instead there were large groups, sharing knowledge and experience and using language to plan survival strategies. The change from life within a threatening hostile environment to a secure family routine has been happening in all social groups throughout the world at different times over the last few thousand years. In those days, thousands of years ago, with our urge to plan ahead, we learnt how to protect ourselves from the dangers of the naturally hostile world. We also began to control some of the natural processes, making fire, and using minerals such as tin for bronze. This can be seen as a grand revolution in control of natural resources which led to the beginning of politics and societies. It meant that for the first time a species interfered consciously with the balance of nature, taking things from it, changing it for its own advantage.

Before Tirefour Broch, without human occupation in far north-west Europe, the ecology comprised stable woodland with grassy glades allowing herbs to grow in the sunshine of natural clearings. Within this steady state mammals, birds and other animals were integrated to form a stable balanced ecosystem. There was no waste. As in all natural systems, the ecosystem was built and controlled from within. It had a kind of peace and harmony. Only external forces changed this delicate balance, and in doing so promoted evolution of large different groups within the plant and animal kingdoms of that changed environment.

There are many sources of evidence for such forcing from outside the planet as well as from within its complex structure. From outside there are asteroid and comet impacts, the gravitational pull of other planets and our moon, the effect of sunspots and solar wind. The God-fearing astronomer Fred Hoyle argued that viruses, and therefore life itself, came from outer space. From inside the planet system, there is ecological succession, erosion from the weather, seasonal environmental change, fights between different species and individuals, and of course, selfish genes.

In the British Isles, the building of Tirefour Broch was one of the first changes to the environment brought about by humans. We are the only species that can force big changes on the environment, and over the last 3,000 years we have been doing that with consequences often beyond our understanding and possibly out of our control.

Imagine the other changes since these early days of the family unit, all caused by the increased human population and activity, grazing, urbanisation and tourism, pollution and waste. They had led to deforestation, species becoming extinct, rising sea level and climate change. Such devastation covers most of the landmasses on the planet, while different but just as destructive changes are being inflicted on the oceans. There are few populated parts of the world which have the same landscape now as 3,000 years ago, untouched by human endeavour.

In the British Isles the only place you can stand and see the same view now as then is on the Burren in County Clare, just south of Galway Bay. This is several square kilometres of limestone pavement which has resisted interference because there has been nothing that could be done with it. Clambering across the huge exposed blocks of rock, grikes criss-cross the surface. These 'dykes', often more than a metre deep, have their own microclimate which attracts a special flora and fauna. The Burren is remote and has a threatening atmosphere, with nowhere to hide from the strong winds. Clouds blow in fresh from the Atlantic. The grey limestone hillsides sweep down to the sea off Black Head, the Aran islands in the distance, still with a small human population of their own and a history not unlike that of the Scottish islands with their brochs and bronze swords.

Some remnants of modern man

Travelling between these ancient monuments of nature in the winter of 2000 brought home with great force the horror of the human impact upon the environment. The local newspaper declared it was 'the wettest weather since records began over a hundred years ago, all brought about by global warming'. There were landslides from new road verges, floods from broken river embankments, fallen trees from old plantations. Land being reclaimed from salt marsh was once again covered by the storming high tides.

It is interesting to note that the damage was restricted to modern features of the environment, artificial landscapes out of balance with the whole system. In contrast, the rocks and soil that had been around for millennia remained intact. These mature structures are part of the enormous system that can survive extreme events because they have been developed within that system through long periods of time. Theirs is a natural peace, a balance within complexity.

We know what the landscape of western Scotland and Ireland was like from the fossil remains that make up the underlying peat. One of the most thorough studies of these environments was conducted by John Birks, a specialist in post-glacial vegetation, on the plants of the Isle of Skye. He reconstructed several different habitats since the last glaciation when the most conspicuous vegetation comprised mixed birch and hazel woodland and shrubby heath. They were grazed by deer, which in turn were kept in check by wolves and foxes. In the cold climate, mammals were best to be smaller, the same species being larger in the warmer south. On Skye, with its bad soil, Birks found pollen and leaves from juniper shrub, while blanket bog and raised bog flourished where water accumulated.

Just as it is shells and bones, the hard parts of animals, that become fossilised, so the lignin in leaves and wood and the resistant walls of pollen grains survive in the suffocating marshes and bogs of these wetlands. Within a few years the fossilised plant parts become peat, then a few million years later, browncoal. With the right temperature and pressure, even greater lengths of time in the hundreds of millions of years turn the organic remains into coal or oil and gas. Pick up a piece of peat between your finger and thumb, throw it into strong acid and then alkali, sieve what's left, and you will retrieve the broken bits of plants and animals from that wetland grave. Specialists like Birks work at the microscope identifying and counting the bits, finally reconstructing the original biosphere and showing how it changes through different scales of time.

Throughout these northern latitudes evidence is left as pine stumps, pollen, plant and animal fragments preserved in the peat. Rannoch Moor is far up in the Scottish Highlands and still has clumps of natural pine, with lots of heather and bog, building up more peat to record the present environmental and biological changes. You can easily see these rare ecosystems from the train, as it passes slowly from Crianlarich to Fort William along some of the most isolated railway in north-west Europe. There is a long climb up around the deforested mountains, slowly pulling onto raised wetlands of bog and marsh isolated from civilisation and even roads. You are in touch with the wild and exposed spirits of Rannoch Moor and its ancient habitat in the vast isolation of the Highlands.

Those spirits of Rannoch create a sense of another age, the clearances of the eighteenth century when landowners started to burn old heather in ten- or twenty-year cycles to rear pheasant and other game for sport, while also clearing the ground for new growths of the same heath. Such fires still maintain the artificial ecosystem on the mountains, holding back mature growth of the trees that would otherwise resume their dominance. The vestigial pines you see around Rannoch are about the only remnants of natural forest in the British Isles. The complete mixed deciduous conifer forest associated with the pines once covered the lowlands down into England and Wales. All that was taken away when humans wanted shelter, ships and sport, and this occurred hundreds of years before the big twentieth-century environmental tragedies that headline the news of environmental decline.

Would the environment have changed in the same way without humans? The answer is a definitive no. That view may be as obvious as many a verdict in a murder trial, but where is the evidence that the jury must have to make a proper decision?

In the nineteenth century when labourers were making the railway across the Rannoch bog, they sliced through sections of peat that had slowly built up through thousands of years. The broken bits of fossil plants and animals preserved in this suffocating sludge tell us what had been happening during the time of their deposition. In the Scottish bogs it is mainly the plants that tell the story. In other places more romantic creatures such as large mammals have left their remains, especially when the sections go back into previous interglacial intervals, about every 100,000 years.

At about the same time as the Scottish railway was being built, other labourers were digging foundations for Nelson's Column in Trafalgar Square. The sides of their trench showed more solidly compressed peat, but samples showed up much the same pollen and leaf fragments as at Rannoch. We call it the Trafalgar Square interglacial, the warm interval from 80,000 to 125,000 years ago, before the last glaciation. Although the vegetation had been very similar to the present inter-glacial, the animals were very different. The labourers were the first to be astonished by the big bones from mammals. The Victorian scientists were the next to be surprised, identifying hippopotamus, an extinct elephant with straight tusks and a weird rhinoceros with a narrow nose. It was hard for Victorian society to believe that in the centre of one of the world's most splendid metropolises once roamed animals that now could be seen only in Regent's Park Zoo, captured and transported from Africa. The realisation sent shockwaves along the spine of human civilisation and culture.

Progress was slow with understanding these amazing discoveries until modern dating techniques could give the changes some relative scale. One difficulty is that much of the fossil evidence from the earlier warm intervals gets scraped away by later glaciers. This makes it very difficult to be sure that the grave has not been moved, or to be precise about the age of what little gets left behind. But from the whole of Europe, North America and Asia, evidence of mammal extinctions during these earlier interglacials is becoming clear.

Furthermore, it appears that the cause of the extinctions was due less to the cold – they could mostly migrate south when it got too cold – than to aggressive hunting by ourselves, *Homo sapiens*. In Africa, new discoveries show that we caused extinctions of almost half the large mammal genera such as giant baboon, three-toed horse, antlered giraffe and many more. In Europe, at the peak of the last glaciation just over 21,000 years ago, we killed off the mammoth, woolly rhino, elk, hyena, lion, bear and tiger.

Scientists and others have been arguing for years about the causes of these ice age extinctions. Several groups with their own special interest have become involved, trying to argue about race and religion, not caring much about the natural processes of change. Some say the large mammals' immune system broke down because of 'hyperdisease' spread by humans. Other scientists see the weather involved with climate change through the ice ages as responsible for the large mammal extinctions.

The aggressive patterns of human behaviour (I dare not call them instincts for lack of clues) appear to have been with us since our inception. Early man migrated out of Africa to central Asia and eastern Europe, then on to China and north-west Europe, and now we are looking for the signs of battle. They are hard to find. Perhaps that's because the relics have been removed by erosion or glaciation, or even because the battles didn't happen. However, there are more and more circumstantial details being discovered to suggest that these early migrants killed off many large mammal species from Neanderthal man to the mammoth. And there are new arguments about the history of man entering North America.

Most scientists now agree that vicious events happened more than 11,000 years ago as the ice was retreating from the last polar glaciations. Then our much less civilised ancestors, Siberia Man, were walking across the newly emerging Bering land bridge from Siberia to North America. They were skilled hunters, and within a few thousand years of their migration 70 per cent of the species of large mammals in North America were extinct.

There is also more and more evidence that extinctions over the last one hundred years result from environmental changes caused by us. As my Taipei tearoom conversation showed, our abuse of the planet is gathering pace and there is little that we do to stop it getting worse, scientifically, socially and politically. There are conventions and summits, a permanent United Nations Conference to look after the

Biodiversity Convention, as well as regular meetings of environment ministers. But the sales of cars and refrigerators soar in India and China; gasoline stays cheap in North America; Europe consumes timber from dwindling tropical rainforest. The tourist industry booms globally. So we go on living through a catastrophe of a kind that our planet has never experienced. It has suffered major environmental changes before, with a consequent loss in biodiversity, but never have they been caused by a selfish species.

Within a few tens of thousands of years we have progressed from scenes of ice age glaciers to temperate interglacials, no doubt causing the extinction of many mammal species. Then there were aggressive hunts by humans for more mammals in Africa, Europe and Asia, and eventually America. Now we humans direct our aggression to abusing our environment. The importance of this sequence of man-made crises becomes clear when you realise that all these human actions have lasted only a few thousands of years, a quick flash in the 400 million years through which our planet has had life on land.

On the other hand, you can argue that these few thousand years are a long time in comparison to the two hundred or so years since the Industrial Revolution. What we've done in our short sojourn on Earth may be comparable to other catastrophic events that happened millions of years ago. It's 65 million years since the last big catastrophe. It happened in a flash. It took seconds for the meteorite to pass through the Earth's atmosphere before the impact explosion and then the long environmental recovery.

Of course, what's happening now is different. The damage we are inflicting has been taking place since the end of the last ice age, about 10,000 years ago, and we are doing it in stages. The most recent stage is dominated by burning oil. Relatively recently, in April 1899, the *New York Herald Tribune* wrote: 'Two motor cars will commence to carry Her Majesty's mails in London itself, the postal authorities having decided to give the new means of locomotion a fair trial. They have quite as great a carrying capacity as the two-horse vans.' Before this, we burnt another fossil fuel, coal, in quantity from the start of industrialisation. Going back further, wood was the main fuel, bringing about large-scale deforestation, and the whole catastrophe began just a little earlier when we hunted so many mammals to extinction. Ten thousand years seems an age by the scale of a human lifetime, but geologically it is only a flash.

Journey from the beginning of time

To understand how this flash fits into geological time, it helps to take a virtual journey on a time machine in which one complete day represents the 65 million years that have passed since the last great environmental catastrophe when the dinosaurs became extinct – a journey in which today's universe is not yet seven months old. Let's go to the beginning of time, the Big Bang. Imagine it happening just after midnight on New Year's Day. Within a fraction of that second, from infinite density, the universe begins. Time begins.

It's intolerably hot and the thick atmosphere makes it difficult to see much. By 18 March of our virtual year the universe has reached 5,000 million years of age. There's not a lot we know about what was happening then – exploding stars, balls of hot gases, no familiar planets to the suns. It's getting cooler, or rather, it's not so horribly hot. By 11 June one of the chunks of matter orbiting our sun breaks into three. Mars and the Earth are formed, and about 500 million years later, on 19 June, our moon. The temperatures gradually fall and by early July life begins on the Earth. (Some say it came from outer space, others that the necessary organic molecules formed from the planet's own inorganic chemistry.) There is no land cool enough for life until around 10 August, so until then early life is aquatic, often with a vast range of, to us, weird forms. Stephen Gould's *Wonderful Life* describes some of these from the Burgess Shale, which Derek Briggs and scientists from the Smithsonian Institution have brilliantly described and illustrated.

Large numbers of these 540-million-year-old fossils from British Columbia show that an enormous structural diversity was present early in the history of life. Because they are among the first non-microscopic organisms they have many unusual features, hard to find in fossils that lived since, and controversy continues to haunt our interpretations. The remains were discovered in 1909 by the paleontologist Charles Walcott, who explained them as 'a sublime conception of God which is furnished by science.' Their different shapes and structures show unusual variety, and many scientists have thought them to be unlike more recent animals that they have seen then representing extinct groups. They also seem to have diversified suddenly and become extinct just as quickly. A more recent approach has been to look at the similarities between the fossils' characters. Links have been made to familiar groups like trilobites and sponges. The confusions should be no surprise, because most new things start off by looking strange.

Our 65-million-years-to-a-day journey is reaching familiar territory now. Through middle August life is evolving very fast, diversifying day by day – vertebrates, ferns, dinosaurs. Some groups become extinct in those early and mid-August days: trilobites and jawless vertebrates. Then, on 20 August, the dinosaurs become extinct as well. On this time scale, that happened yesterday.

On the chime of midnight for the start of this new day, 21 August, we wake from one of the planet's most horrific nightmares. The northern hemisphere is completely blacked out with smoke and dust in the first milliseconds of our virtual day. The vegetation returns to normal after a few minutes, with only a few changes, while the oceans take longer to clear up the debris that has rained down from the dirty clouds. It has mixed into the sea, and robbed it of much oxygen, causing extensive extinctions of plankton and fish.

At an hour past midnight the noxious outfall from the nightmare has completely cleared and a bright new world is beginning to take advantage of the new opportunities. It is not unlike the upturn of the Western economies after the Second World War – hesitation to reacclimatise at the beginning, then a surge in diversification to reach new highs. This is a pattern I return to in chapter 5. Throughout this Paleocene period of geological time, up to about 3 am, the environment is establishing new ecological niches in the very warm climates. This allows the number of mammal species to peak by the 6 o'clock dawn, comfortable in the new reality that they are no longer the prey of dinosaurs.

A very clear trend is developing which we will see characterising this whole last day of our journey. There are many new species ranked together in new large groups of genera and Families, but there are very few extinctions. Overall, there is a massive increase in biodiversity. In the early morning there are the first primates, the first horses, the first whales, whole new major groups of animals, each with hundreds of new species.

At 9 o'clock in the morning it is 49 million years ago, during the period of geological time we call the Eocene (see figure 1.2). The planet is becoming quite a familiar place, with dense mixed woodlands and savanna grazed by herds, and there are even cocks crowing. Monkeys are one of the big new groups to originate and quickly diversify. Global temperature differences are much less than today's, the tropics being about the same and the poles equably temperate. The shores of the Arctic Ocean, as well as the hills of the Antarctic continent, have warm tranquil climes with low sun in the long summer and little if any frost in the darkness of the long winter. If there had been humans and a travel industry then, it would have been a tourist's paradise for half the year.

Just before lunchtime, 35 million years ago on the geological timescale, temperatures have peaked due to high carbon dioxide concentrations in the atmosphere. The greenhouse effect in these times is much stronger than now. Continental drift causes the North Atlantic to widen and at about the same time to open into the then temperate Arctic Ocean. Temperatures at the poles begin to fall. We're not sure how this happened but perhaps it was caused by some astronomical phenomenon, or by the changing positions of the continents and oceans.

In the nowadays highly populated northern temperate regions of the world, tropical rainforest stretches from Asia, eastwards to Europe and the newly separated continent of North America. The very rich faunas and floras begin to take on a familiar look. The world is becoming

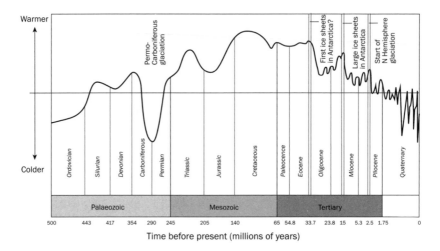

Fig. 1.2 The Phanerozoic timescale of the last 500 million years and its classification into Eras and Periods. The curve shows changing temperatures through this time. (based on W. B. Harland *et al.* 1990 and S. Lamb & D. Sington, 1998)

more varied, with more species than ever before, and its ecology more diverse, with a wide variety of modern habitats. With the wider-ranging weather and climate, and the peak diversification of animals and plants, come maximum complexity and range.

By now it's early afternoon and many familiar groups of animals and plants are making their first appearance. While only the odd species is becoming extinct, all Family groups continue to diversify. There are now apes in the warm forests, and other new Families including camels and deer in the shrub. There are also new grasslands, legumes and palms. Evolution continues at a high rate, though as yet, in the Early Miocene, there are few, if any, modern species. They will come later, in the evening, migrating south to the more restricted equatorial latitudes of the modern tropics, and evolving into different groups to populate the cooler forests, wetlands and grasslands of the more temperate regions.

Through the afternoon, temperatures become much lower, the weather much more extreme, the ocean currents developing trends familiar today. The changes are not smooth or simple, but occur as oscillations over different timescales. Present-day changes in weather show that these uncertain variations continue. Around teatime the poles become covered with permanent ice, which slowly spread towards the Equator, developing glaciers and icebergs cooling the oceans. Cold-loving mammals grow large woolly coats and take advantage of the new niches in the freezing landscape. They diversify accordingly.

Early hominids first appear around 11 o'clock in the evening, and at 20 minutes to midnight *Homo habilis* is soon followed by *Homo erectus*. Neanderthals and then modern humans originate a few minutes before 12 o'clock, depending on your view about the precise time, place and definitions of the species. This happens during the Pleistocene, the period of the ice ages, which began nearly 2,000,000 years ago. The northern hemisphere continents become glaciated at five- to ten-minute intervals for half an hour or so before midnight, the end of our day. But then, at just two seconds to midnight Jesus Christ is born to begin the first millennium of the calendar we record today.

Into the future, no one's too sure how long our virtual journey will last, how far away that elusive end-point really is. Even the dates I've given for the virtual journey so far are debatable, different specialists arguing about them all. But what does seem to be pretty well agreed now is the basic plot of the story. Our Earth really is on this kind of journey. These changes really are happening.

Our planet isn't the only part of the system that is following the sequence of origin, expansion, maximum diversity, then finally contraction. Species and groups of species do the same thing. I suspect that man-made institutions like governments and empires, businesses and fashion also follow the same patterns, but I have no data to test that hypothesis. If so, this raises an important universal issue: can all these apparently different systems be following a similar pattern?

But so many different things are influencing the Earth system itself that it's hard for us to be sure of the full effect of each one. We are just beginning to be able to take those that we recognise together, only just beginning to think of our own role in the system of nature and the many different influences we are making. Can the behaviour of our ancestors building Tirefour Broch 2,100 years ago, or hunting mammals in Alaska 8,000 years ago, be seen as part of the same evolutionary process that I've painted for the last 65 million years? Even more basically, can the human mind, from our position inside the system, have the ability to interpret these complex patterns and explain the way life works on the whole Earth?

Challenges for a young research group

I think answers to this question will come from the interdisciplinary revolution that's only just beginning. Coincidentally, information technology is making it possible to join together data from different disciplines. What is starting to happen for the first time is clearly shown in the composition of my research group. I'm a fifty-nine-year-old paleontologist, working with two thirty-year-old computer buffs with pony-tails, a young woman who knows more about biodiversity websites than anyone else, and Dilshat Hewzulla, a young mathematician from China who's a genius at analysing large datasets.

We're all from different backgrounds, working on the same problem: the changes in biology and environment through geological time. We all have very different knowledge, different skills, and different tolerances of computing. We are all totally dependent on one another. It's unlikely that our group could happen in a large mainstream university department because all we have in common is individuality and eccentricity. We came together by accident rather than design, yet these oddities bind us together.

Dilshat has introduced to our group another young computer scientist from Urumqi, Alim Ahat, a mathematics graduate and director of a new software company in that city, Ugarsoft. It's the leading computer company in Xinjiang, a province of China with 56 million inhabitants. There is also my old colleague and friend Richard Hubbard, who cycles round London wearing sandals, brightly coloured trousers that he makes himself, an Aertex shirt and keys around his neck. Whenever the hot nights of the Proms were televised it was usual to see him standing in the front row. At Oxford he studied chemistry, then archeology at London, and now he has an international reputation for performing principal components analysis on our paleontological data.

We work together in a simple and logical way, each taking responsibility for our own expertise and all coming together at the end to use our different perspectives and common sense to make an interpretation. I start off the process by finding new data from the scientific literature or the internet. That is then validated, cleaned up and cropped, maybe as much as half being thrown away. As I will keep mentioning in this book, the fossil record is notoriously poor, with gaps, uncertainties and much that is plainly incorrect. Mathematics and statistics help sort it out. Others in the group assemble the cleaner data into spreadsheets, write programs to compare them with things searched from other databases, and compile methods to analyse and model. You can see some of this work at *http://www.biodiversity.org.uk*

The changes in information technology and data availability are happening so quickly that we come to accept a danger that the work will be out of date before it's finished. Another challenge is our bid to compare and integrate data and concepts from mathematics, physics, chemistry, genetics, evolutionary and systematic biology, and cognitive psychology. This is bound to lead to new ways of thinking about what environmental and evolutionary processes do when they are at work on our planet.

This holistic view shows us what lies between the extremes of physical and biological change and teaches us that physics and biology work very differently. One has Laws, the other doesn't. One can be described quantitatively, the other qualitatively. A question is whether these extremes can be compared, whether physics and biology can be understood and described in the same way. This is more than a semantic issue, because we need some way to monitor and conserve the changes humans are inflicting on the stock of nature. I fear that the loss in biodiversity, whether it be ecological, botanical, zoological or genetic, seems to be inevitable, whether we count it or not.

The word 'biodiversity' sprang into use from the 'National Forum on BioDiversity' organised by the US Academy of Sciences in 1986. It was a major topic title at the 1992 Rio Congress which begat 'Riodiversity' and more reasonably, biodiversity. It is an interdisciplinary concept, enabling comparisons of previously separate ideas, a new way of thinking about biological systems. The biggest records of changes in biodiversity come from Europe and give an idea of how the problems of loss are being approached by scientists, industry and politicians. Last century, Europe lost most of its sea mammals, natural forests, grasslands and many other habitats and species. Other losses are high when measured in terms of local abundance, but the same species show relatively little change when expressed as regional diversity. It all depends on how you present the figures. I prefer to rely a little on feelings and my common sense.

Another quantitative estimate of the new century makes a chilling comparison to this European observation. It comes from a recent study of 'Who will feed China?', where 1.2 billion people now live, and makes cheeky comparisons between East and West. If every Chinese ate just one extra grain-fed chicken a year, that would account for Canada's annual grain harvest. If Chinese used motor cars the way Americans do, global oil output and CO_2 pollution would both be more than doubled.

But the huge complexity of what's going on does lead some of us to broader views of evolutionary processes, helping us to better understand the living systems on our whole planet. From the stimulus of the 1992 UN Convention on Biological Diversity, now ratified by over a hundred countries, these factors spread over into groups asking far-reaching questions concerning what science can do to help. One major job is to monitor biodiversity. Another is to educate the global population to respect the planet. Geneticists and pharmacologists are busy defining and extracting beneficial chemicals from threatened plants and animals. Researchers in agriculture and horticulture need the full genetic stock to breed new varieties. Scientists can also advise politicians and administrators and global companies, but usually these people don't want to listen.

Monitoring this delicate biodiversity and keeping track of disappearing species from different places is proving to be harder to organise than you would think. Just after the Rio Convention several national and international groups began to plan monitoring projects. The first, Species 2000, began in the mid-1990s and aimed by that date to link together lists of all known animal and plant species. Our plan was to make internet links to the world's authoritative biodiversity databases. We attracted the involvement of fish specialists in the Philippines, the insect group at the Smithsonian Institution in Washington DC, viruses in Tokyo, legumes in Southampton and fossils with my group. David Gee, one of the pony-tailed anoraks in my group, devised a program to query these at once, from one request. You can do this now from *http://www.species2000.org*.

But after five years' work, the target is still very far off. Species 2000 and other projects have involved a lot of talking, a lot of travelling around the world going to meetings, and a lot of disagreement about which standards to use. They also got caught in a Catch-22 situation, which may mean that they can never succeed. The scientists involved believe strongly in precise objective monitoring; because that's so slow and expensive, nothing happens very quickly. When private-sector money is offered, the condition is for fast results and common names, none of the precision of whether it is one species or another. Most scientists run away from this kind of thing, scared of the commercial sector and of not winning a good reputation in the one they know.

One monitoring project that does seem to be working began long before the Rio Conference. It's run by UNESCO's conservation group and is called 'Man and the Biosphere', MAB. The programme has devised an international system to conserve particular biosphere reserves, natural areas with animals and plants that can be sustained. Now there are 368 such areas, in 91 countries, and the number is growing. They are selected, monitored and maintained by a whole variety of environmental groups and agencies. In the UK, some are rather formal government committees, others anorak-clad bird watchers, and all contribute to a huge database (*www.biodiversity.org.uk*).

Another approach to monitoring biodiversity has been formulated by a group of scientists from Oxford and Washington DC. They have identified a number of hotspots, parts of the world with high diversity in exceptional danger of loss. Hotspots such as those around the Mediterranean are estimated to contain 44 per cent of the world's 300,000 known vascular plant species and 35 per cent of the total known species of mammals (4,809 species in the hotspots), birds (9,881), reptiles (7,828) and amphibians (4,780). But how can we monitor these very sensitive places and police them for restoration to even something approaching their former state?

Data like those from these hotspots are accumulating at an increasing rate. Money and expertise are coming from national and international agencies to survey the biodiversity of these special regions and make it available publicly. These early schemes are scattered around different agencies and have different standards and objectives, but eventually they will come together. The internet will see to that. During the Rio Conference it was clear there wasn't enough information to support the arguments of gloom and doom that were being expressed publicly. Now the situation is quite different, though a lot more has to be done. Around the time of the Rio meeting, my own research group started to build databases of where fossils were found, not realising where the work was going to lead. We thought we were making a catalogue of species, detailing their geological age and where they have been found. It's still far from complete.

Instead, our energies have taken us into the very different world of data analysis, the mathematics of complex systems and to the edge of chaos theory. Our approach is to standardise all the data into the same format, with separate Microsoft Excel columns for names, ages, location, ecosystem and other variables. To make sense of the incredible amounts of such data, we propose models against which to test those data. If we think biodiversity is changing in a particular way, we describe that way with a mathematical equation and see if the data can fit it. We test to discover if there are any broad trends showing up to conform to the model. To our great surprise the patterns that are emerging from our analysis of records of extinct plants and animals are clear and definite, and our scientific results confirm our right to be very worried about what is happening to life on our planet.

We have found consistent patterns in these evolutionary changes, in groups of animals that are extinct, and in others that survive. The changes follow a simple model that can be expressed as a mathematical equation, and we use this to predict likely trends in evolutionary change. It's rather like how weather forecasters accumulate data from earlier records of location, temperature, wind and pressure. The patterns are then used to calculate how the values will go forwards in time, and separate statistical methods give a reasonable amount of certainty. We have been doing something very similar working from our evolutionary patterns, and it's now very clear that there is sudden and unexpected interference in the patterns: the environmental changes caused by man.

Darwin's mentor, Charles Lyell, was one of the first geologists and is best remembered today for his maxim 'the present is the key to the past'. This principle urges geologists to interpret ancient structures by observing the way things happen in the present. I fear this over-simplification has misguided many innocent students of geology, as journalistic phrases often do. I will argue in this book for another way to explain the urgent crisis for biodiversity. This is by inverting Lyell's phrase to become *the past is the key to the present*. That's what our article was all about, the one Dilshat's e-mail to Taiwan told me had been accepted for publication: computer modelling from fossil data. This paper works more subjectively than most modern evolutionary theory. That is one reason why the work is so controversial. In it we interpret evolutionary patterns in the fossil record with the statistics and mathematics of complex systems and chaos theory. As I will explain in chapter 3, we compare our results with those from three other well-worked sets of data: one set derived from purely random processes, a second from artificial sources, and the third from natural ones. Our results have the same pattern as those from the third system. Nature, we argue, is in control of itself: biological evolution is controlled from within that system of life on Earth.

Our analysis encourages a different way of looking at the large datasets of biological information from all the different disciplines now being brought together. It takes us away from the search for laws, precise definitions and quantitative testing. Instead, it leads us back into the state of mind where we worked with mystery and uncertainty, accepting that we cannot explain everything in detail. The systems of life in the diverse and changing environments on this planet are so complex that only subjective methods can assess them now. Scientists need to accept life's beauty, and work with it like a fairy story, changing the focus to fit the particular needs of the particular circumstances at different times. Narratives change. In the past science has often responded to what it knew by telling stories about the world and finding the facts to confirm them. Today, the facts are telling a new story, less welcome than some of the earlier ones, and with an ending that we may not be able to change.

2

Extinction

Jurassic story

Children have always enjoyed fairy stories. Their early imagination conjures fantasies that make the spine tingle yet leave them safe within the protection of the parental storyteller. The contrasts between good and bad couldn't be clearer and the eventual triumphs of right over wrong are an inevitability. The importance of these perpetual fairy stories as metaphors of human thought is well recognised by psychologists, and there appear to be complex reasons to account for their continuing popularity in the face of so much modern competition.

As in *Jurassic Park* and *Walking with Dinosaurs*, the images of the land of the dinosaurs are frightening, yet the protection of knowing they are virtual images is comforting – like Little Red Riding Hood's wolf, they aren't around any more. They are no longer a direct physical threat but remain as frightening images in our minds.

Imagine the scenery while striding along the quayside at Lyme Regis in Dorset, the location of *The French Lieutenant's Woman* and countless television films about eighteenth-century swashbuckling. Centuries ago, the curve of the quay was built of large blocks of local rock which still protect the small harbour from the weather of the English Channel. The stones show remnants of earlier life embedded in the rock, fossil mussels and oysters from the Early Jurassic sea that flowed along this coast 200 million years ago. We walk down the quaint old steps, slipping on familiar slimy green seaweeds, ancient algae which also grew in the Jurassic sea. Down at the bottom of the steps the water laps as it has almost always done. We climb into our small rowing boat and the oarsman unties the ropes to cast off. Out of the small harbour and into the bay we float to a world apart: miles of ocean, a warm wind. The boat rocks with the waves and transports us to the edge of fantasy. We are off to see the dinosaurs.

Two hundred million years ago the sea there was much warmer than today, with very little wind and only small waves. Visitors like us would find breathing difficult with less oxygen in the atmosphere, and with the high humidity we would feel distinctly uncomfortable. We cool off many times by swimming from the boat in water much less salty than what we are used to. One of our group swims into a large submerged object shaped like a circular buoy, which then squirts sea water all over us in the boat as we row off quickly, scared by the turbulence. It is an ammonite.

These predecessors of the modern nautilus have a flat spiral body protected by a tough round shell. They move sluggishly through the waves, just a little faster than us. Some of them are about the same size and just as seaworthy as our boat. Their large buoyancy chambers under the carapace suck and blow air like a jet engine, and their mouths flush out fish and plankton with clumsy movements. From evidence in their fossil graves we think that many of the smaller species did not dare leave the shore, fearful of being chased by larger predators. They basked in the strong sunshine waiting until the high tide for their next meal. Many of the fish in the same sea have long sharp snouts specially designed for fighting, while in the air there are the flying monsters, which we can see from our exposed position on the rowing boat.

A flock of more than twenty skin-winged pterosaurs swoop down to attack, their spearlike teeth ripping into the surface-feeding fish that we have attracted around our boat. The sea turns dark as the energy of these beasts churns up the water to rock our boat. But they are as scared of us as we are of them, and they fly off with their prey as fast as they arrived, leaving behind a trail of debris. That draws new attention to our terrifying position, this time from feathered foe: scavenging birds and *Archaeopteryx*. Their visit is as timid as the ammonites'.

These early ancestors of the birds have left evidence in the rocks that they swallowed the remnants whole and then regurgitated the indigestible remains, as owls do today.

There is also evidence of the food chain to show that the ammonites ate the plankton, that fish ate the ammonites, and that dinosaurs ate the ammonites and the fish as well. Through the earlier Triassic Period, sharp-toothed dinosaurs called nothosaurs, 4m long with small heads, long necks and tails, swam with small paddle-like limbs and ate the fish. We know that fish ate ammonites because their teeth marks have been found on ammonite shell tests. Sometimes the food chain was extensive and new or different species got involved, arthropods for example (invertebrates with segmented bodies, like lobsters). Broken bits of their bodies have been found in the fossilised dung and stomachs of marine reptiles as well as the regurgitated pellets from birds of prey. In their turn, the arthropods may have fed on small ammonite species, which then lived on the plankton ooze.

The food chain is part of the system, all changing slowly in tune with the environment. Together they form part of a gentle rhythm. If one part changes, the rest is affected. Nowadays, very similar rhythms are generated when crowds of people create an atmosphere of bustle of the kind you sense in airports and hotel lobbies. Out of the season there is a constant stream of people walking in different directions, positive with a sense of purpose, living and working with quiet efficiency. However, around Christmas and the high summer holidays they are different places, overcrowded beyond what the system was designed to carry, with people jammed, waiting, arguing in frustration.

The pace of life on the planet also changes through time. Ice ages and periods of high volcanicity are the equivalents of the busy, rushed days before Christmas at Heathrow. Conversely, the Jurassic and Cretaceous were off-peak times, with only one or two spurts of intense activity. Throughout this time biological evolution continued at a slow and steady pace in response to the equally modest environmental changes. Just as in airports and hotel lobbies outside rush hours, there were a few new journeys started, a few new species registering, but no major upsets. The relatively stable environment, the ecological balance and the steady growth in diversity saw evolution work mainly at the species and genus level. It was a bustling rhythm but without major divergences or catastrophes.

Back in our Jurassic boat we approach the shore, where we can see species of the two dinosaur groups. The Saurischia, with hips like lizards, stand on two legs to fight other animals in the famous *Tyrannosaurus* pose. Those with the heavier bodies and smaller heads, the Ornithischia, browse on the tough leaves of cycads and conifers, and walk around passively on four legs. This other group, with hips like birds, were vegetarians. They were also heavily armoured against attack from early Saurischia like *Tyrannosaurus*.

We know so little about their inter-relationships that the view from the rowing boat remains a fairy story. A popular view is that all this fighting, all this competition between individuals and species, is the motor of evolution. That is a myth from Victoriana, placed under the Darwinian banner of 'survival of the fittest'. It's an old-fashioned concept that should be banished to the annals of what is wrong about biology. Now, we know that the complex relationship between the organisms and the environment is also important. Evolution is less to do with winning battles between species and individuals, more to do with being able to live well together in the same environment. It is not necessarily the strongest that succeeds, but the most adaptable to new environments that might develop suddenly and unexpectedly.

In the tranquil times of the Jurassic and Cretaceous there were very few and undramatic environmental changes. Temperature and CO_2 concentrations steadily increased well above today's values. The vicious battles between individuals and groups of Mesozoic monsters did not encourage major evolutionary changes. New species took over from earlier ones, a few new Families originated when there was a major altercation in battle with other animals or with any of the rare environmental changes. A few species and even genera became extinct. There was peace and relative quietness on Earth: evolution happened on a small scale, origins mainly at the species level, a few genera and fewer Families. Without big environmental changes there are few, if any, big evolutionary advances. Especially during the middle of the Jurassic there were only small and subtle changes in the marine and terrestrial environments. Without catastrophe there were only small evolutionary changes during the time, usually at the level of the species and genus.

Of the many important things to be learnt from these most tranquil of ages, there is one that most people do not expect. A popular view is that all the fighting, all the business of one thing eating up another, is the primary drive of evolution. They say it leads to the evolution of man and our seeing ourselves as the most powerful beings, sitting at the top of the evolutionary tree. This is not how nature works. The ammonites that ate most fish or resisted attacks from a soaring *Pteranodon*'s beak didn't necessarily do any better than the more compromising species. So the bravest ammonites, charging off to battle in the front lines, perished in larger numbers than the more modest cowards who had found a safe niche.

What did survive were those most able to succeed when the environment changed. So the creatures that come to dominate at any given moment do so, not by power of fighting but by chance. They have just happened to fit into new surroundings at that particular time better than the others. As the environment, or internal biology, or social behaviour, changed, so they just happened to be in the right place at the right time with the right kind of biology. Now, humans think we are at the peak, just as the dinosaurs were through these Mesozoic times before the Cretaceous-Tertiary mass extinction. But once again the environment is changing dramatically.

Geologists are the supremos at understanding environmental change. In the formations of rocks they recognise signs of change in the atmosphere, the land and the sea, all connected and dependent on one another, linking events through time. Among the more dramatic changes through the Earth's history has been the rise and fall in sea level – affecting drifting continents, changing climate and weather systems – different atmospheres, and changes in sea composition. The planet is still clothed in very many environments which are changing, even now, though at different rates. It is an incredibly complex system, which we are only just beginning to follow. Our Jurassic boat trip, offshore from Lyme Regis, can help with some observations of the kinds of thing that happen during hundreds of millions of years.

As well as its quay, Lyme Regis is famous for its fossil ammonites. Theirs was a very different sea to the present English Channel which laps up to the Dorset coast; indeed, it was quite different to any sea in today's world. At the beginning of the Jurassic period, all the world's land was one huge C-shaped continent called Pangaea (see figure 2.1). What is now southern England was low-lying land right inside the concavity of the C, the site of river deltas from the north, west and south. In this very sheltered coastline the shallow sea carried lots of debris from the land. This meant that the sea was brackish, with much less salt than usual. It was ideal territory for ammonites and dinosaurs, as well as marshy plants, but not good for many others.

Slowly, the continental plates in that region began to move, causing Pangaea to break up from the central axis of the C-shaped land mass. The sea level fluctuated, the dry landmasses began to form, the marshes grew less brackish, and eventually a new shallow sea stretched westwards to the other side of that C-shaped continent. The complex interaction of processes swung backwards and forwards, but within a few million years North America had separated from South America. In turn, this caused global sea currents to change completely and rougher weather came to Lyme Regis as the sea level rose. You can see this changing global geography at the website (*biodiversity.org.uk/maps/palaeo*).

These complex pictures contain objects from many different disciplines. To make a sensible interpretation the trick is to work out how to separate all these oscillating signals from different parts of the system. It needs full and broad knowledge and often demands evidence through long periods of geological time. In the first place we have to recognise the changes by breaking the system down into its component parts. Then we have to put these facts back together again to prove that they really are working together as part of Earth's whole system. At last,

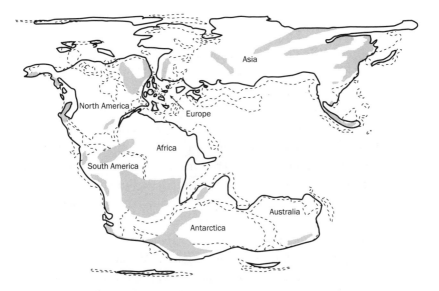

Fig. 2.1 Palaeogeographic world map for the Early Jurassic, 200 million years ago. Ancient coastlines are shown with a solid line. The dotted lines show the present continents. Upland areas are stippled. (after R. N. L. B. Hubbard & M. C. Boulter 1997, *Palaeontology* 40, 43–70)

that trick is beginning to make some sense through a multi-disciplinary scientific approach.

Without clear evidence, there is a suspicion from the foggy images of the whole Earth system that nature changes very often. These changes can take anything from a matter of seconds to millions of years to occur, at different oscillations and cycles. In Lyme Regis, 200 million years ago, the evidence points to many different cycles of change for many different environmental factors. Although the ecological changes were on a small scale throughout the Jurassic, there were enough to stimulate small evolutionary changes.

Biodiversity is a complex system and was growing even then, with off-peak rhythms of change. The continents were drifting, food cycles were changed by slight shifts in climate and ecology, CO_2 concentration and temperatures were rising. But it was so long ago that the evidence is distorted, fragmented or often destroyed by erosion. It's difficult and often impossible to understand the timescales of the changes.

When you are suddenly dumped into the middle of changing systems like these it's hard to get your bearings. For example, our understanding of long-term changes in the weather depends on whether we have data about the changes over a broad sweep of space and time. No wonder it's difficult to say what life was like 200 million years ago on the basis of describing the Jurassic rocks at Lyme Bay, especially since all the changes appear to have been relatively modest. But some ideas about how to make sense of complex systems came 150 years ago from a surprising source: a retired railway engineer.

Herbert Spencer stopped working on the railway at Derby in the 1850s to be a successful writer and thinker. At one time he was the deputy editor of the *Economist* magazine. His social theory was that the best-adapted humans reproduce most effectively. 'The survival of the fittest' was his phrase, enjoyed by Darwin and ever since by politicians and misguided students. In the 1950s, the famous American geneticist Sewall Wright applied the same ideas to animals in the wild, seeing them as 'fitness landscapes'. Wright was one of the first to try to monitor genetic and evolutionary changes, along with J. B. S. Haldane, George Gaylord Simpson and many other bastions of evolutionary biology in the mid-twentieth century. They didn't have help from DNA or from much environmental data but they did have ideas about processes. In population genetics the conventional variable for fitness is 'W', encouraging one of his students to ask Wright if that 'W' was after 'Wright'. 'No,' came the reply, 'it's for "Worth".'

The rugged peaks of a fitness landscape mimic the conflicts and constraints of biodiversity (see figure 2.2). Both have features that show a range of sophistication: good at this and bad at that, works well in one state and not in another – different certificates of fitness. The complex system of biodiversity, whether in the tranquil Jurassic or the fast-changing world of today, has the fittest individuals at the peaks where they are most likely to succeed.

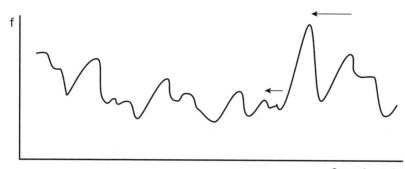

f

Genetic code

Fig. 2.2 The fitness, or ability of species to survive, is plotted against genetic complexity, as a theoretical concept. To improve their fitness, species located in the troughs have smaller barriers to cross than species on the peaks. The fitness landscape reflects the ease with which evolution can proceed for different species at different times. (after P. Bak, 1996)

What's new and exciting about fitness landscapes is that some of the oscillations we find in the fossil record follow the same patterns that we find in many other natural systems. Just as natural landscapes can cope with the complexities of changing weather and ecology, so all features of life on the planet fit together systematically as they keep changing. Organisms at the peaks one moment slip down when some other part of the system changes, but equilibrium is maintained and the rhythms continue.

Right through geological time internal and external planetary factors have made enormous impacts on land shape and climate. Internal factors come largely from plate tectonics, external ones from things like impacts. It has been rare for the environment to stay the same for very long, because the system is so dynamic that a kind of chain reaction goes on in different directions and dimensions. At the time of the ammonites and dinosaurs, during the Triassic and Jurassic periods around Lyme Regis, the temperature and climate were increasing in very slow cycles. Within these complex ups and downs different sediments around the Dorset coast hold clues to what actually happened. Evidence of the changes in vegetation was left behind in the rocks as fossil pollen and spores from conifer and ferns. They show a steady rise in temperature then, which fits with increases in coral reefs, bivalves and many other groups of shallow-water marine organisms which enjoyed the warmth.

Not all scientists trying to reconstruct changes in the Jurassic climate agree with this account. For example, results from our own research group are in direct conflict with those from two other students of Bill Chaloner, Jenny McElwain and David Beerling. Our group's work was done mainly by Richard Hubbard, who analysed counts of many thousands of pollen and spores from Triassic and Jurassic sediments, 210–190 million years old. One of his principal components was a group of plants we believe to be cold-loving, and they show up right on the boundary between the Triassic and Jurassic periods, 205 million years ago. Jenny McElwain and David Beerling found a warm phase at just that same interval. They had assumed that the density of pores on 205-million-year-old fossil leaves is a marker for CO_2 concentration. Fewer pores mean that less water is lost so temperatures are higher. We'll have to wait for evidence from elsewhere to see who is right, whether it was a cold snap or a warm one.

Disagreements such as this are scattered through the scientific literature and occupy much time in tearooms and at conferences. The whole scientific way of thinking is based on the challenge of proving something wrong, on refusing to accept the conclusions of others and hopefully being able to prove the hunch right. Journalists and teachers do a bad job in conveying this conflict to members of the public, let alone to politicians. Both groups want straightforward answers to straightforward questions and don't understand it when they can't get them. Invariably they force out an unsatisfactory compromise from the bullied scientist and give ill-informed impressions about what's going on. AIDS, BSE, GM food and foot and mouth disease are recent examples of confused public awareness of really complex controversies which, as yet, have no clear or complete explanations.

In the Jurassic landscape, animal and plant life was well adapted to the rising temperatures. It meant that many of the features familiar in our presently much cooler climate were missing. One is fur, which then would have made mammals' bodies too hot. The thin leaves of deciduous trees and herbs would be scorched, so they were not evolved. The few small mammals that were beginning to evolve had no fur and the plants remained thick-skinned. As with other new groups the diversity and numbers of individual mammals were small, a measure of their limited power to compete for space and food with the other beasts. They were on the lower peaks of the fitness landscape. It was to take major changes in the environment before they were free to diversify and grow larger.

Today, the beaches and the cliffs of Dorset are full of the fossilised remains of these times, especially the dinosaurs and ammonites. You can also see Jurassic coral reefs preserved at the base of Church Cliffs, relics of the shallow Jurassic sea, and, as the sea deepened, so limestone and shales form the upper part of the cliff to bear witness to marine sedimentation. The majesty of famous Dorset coast cliff scenery comes from the series of small environmental changes like these. Other ancient events show up as bands of pebbles, upright yellow sandstone, slouching grey clay, folded strata, caves and bridges eroded out of softer rocks by the sea. These forms of geomorphology give further clues to the Jurassic landscape and how what's preserved has been altered time and again since they were formed up to 200 million years ago.

Westward Ho!

The variations in geomorphology come from changes in weather and climate, in the position of the continents, sea level and so much more. These factors, in turn, influence the pace of evolutionary change of the organisms in the same system. As with so much of nature, their world was knitted together into a humdrum rhythm of normality. In different ways all organisms reach this state of dependence on others, both other species and other members of their own. There is also a dependence on the particular environment for each individual and community. The eventual effect of this interdependence is for organisms living together to establish a state in which there is a more or less constant harmony. Together, the whole of biodiversity reaches a stable state. The mature systems appear to take care of themselves and don't seem to change unless there's some interference. Individual organisms become fully integrated into the community of their own species and into their wider ecosystem. A question we have to ask is whether or not such stable states can go on for ever, or whether 'system Earth' makes sure that, in time, some kind of change is inevitable.

I used to travel regularly from London to Devon, relaxing through the English scenery of Brunel's Great Western Railway. It goes west from London to Bristol and then south to Exeter and Cornwall. The railway was one of the first of its kind, built during the 1840s just after Darwin had returned from his expedition to South America on the *Beagle*. Taking this train journey through the railway cuttings of western England is also a geological voyage. It follows the strata of rocks back through the sequence of geological time (see figure 1.2).

The clays of the Thames valley outside London were sedimented from shallow seas in the first half of Tertiary time. Heavy grey clays alternate with lighter sands, all getting older as we travel west. They are 55 million years old at Reading and 5 million years older at Newbury. An hour from London, the train passes out of this modest landscape into the grand sweep of the chalk downs through quaint market towns like Hungerford and Marlborough, past the white horse on the chalk hill near Westbury. Jurassic sandstone cottages show up around Bath. They pass over the 200-million-year-old rocks around the redbrick terraced houses at Taunton. It is the same sequence of rock formations that outcrops on the Dorset coast just to the south.

As the train passes the changing environments, the different local building stones still serve as reliable clues to what was sedimented 200 million years before. There are few parts of the world where it is possible to travel back so far into geological time so quickly and see the sequence of rocks as they were laid down in the changing environments of the past. The place on the planet now occupied by the British Isles was in a pivotal position through most of the Mesozoic. That spot was inside the C shape of Pangaea (see figure 2.1) before it split up. Sometimes it was land, sometimes sea, and often on the floodplain between the two. Placed between Euro-Asia and what is now North America, the region is a microcosm of the despositional history the two blocks experienced on a larger scale.

The train stops at Exeter, on extremely complex geological structures due to the prolonged influence of the Dartmoor granite nearby to the west. Potassium-argon dating shows that the granite comes from volcanic eruptions 280 million years ago (*www.phdcsm.freeserve. co.uk/overview.htm*). Dartmoor itself, as well as Bodmin Moor and Land's End to the west, are the remains of hard granite from these volcanic structures of the early Permian (see figure 1.2). The region around Exeter and strips of land down to Torquay and due west to Crediton are what remains of the lava flow from the eruptions. The famous red soils of Devon are even more widespread signs of the iron-rich rocks formed more than 200 million years ago. In good scientific tradition, the age of these events has been in dispute for twenty or more years, as geophysicists refine their dating techniques.

As usual, one morning on that train I had some papers with me. On this occasion I was sifting through some graphs of results from a new search of our fossil occurrence database. Funnily enough the data were from fossil records in rocks showing the very same range in geological ages as the train journey: from the tropical swamps of the Tertiary from London to Reading, the deep oceans of the chalk near Hungerford, the crisp foliage and deserts of the Jurassic near Bath, to the swamp and ocean of the Carboniferous in North Devon. The graphs condensed data from all over the world, not just south-west England, and they were a horribly confused and chaotic mess. They plotted the number of records of each kind of fossil in every one-million-year interval of time. All my earlier attempts to interpret them had failed, so I had put them to one side, waiting for a quiet moment.

Most of the fossil records were from plants and microscopic marine plankton, long names, hard to pronounce, showing up the changes in vegetation of past forests and seas. They were from sediments from all ages back to 200 million years ago. On land, at first ferns and exotic evergreen conifers and cycads dominated the flora, and then flowering plants became common in the lower-lying land. In the sea, plankton blooms dominated most of the warm water, though through one interval they were conspicuous by their absence. The environmental changes of a catastrophic event seemed to be showing through.

Somewhere near Westbury – I can't remember exactly where because I had become so excited – I noticed that just five of the hundreds of curves showed the same clear pattern of change. The changes also showed up on the five graphs at about the same time in the Cretaceous rocks, 90 million years ago, at the Cenomanian–Turonian (C–T) boundary. Each curve plotted the number of occurrences through geological time of the plant groups I had selected and all five suddenly rose at this same time. The groups touched on the most sensitive mystery of paleobotany: how did the flowering plants first evolve?

Surprisingly, birch and elm were among these first five to diversify suddenly then. There was also a rise in the records of an early mangrove plant. But the biggest rises were of two large extinct groups of dispersed pollen, the aquilapolles and the normapolles. They have been known for about fifty years and look quite different from any other pollen, fossil or modern. Their occurrences peaked 20 million years later and then slowly became extinct. We still don't know anything about their parent plants' structure or status but they do appear to have played an important role in early flowering plant evolution.

Why, you might ask, should I be so excited about a few graphs? For years my fellow scientists have been trying to solve what Darwin called the 'abominable mystery' of the origin and early evolution of the angiosperms, flowering plants. Many new discoveries of beautifully preserved flowers have pushed the date of the earliest known flower further back in time to the beginning of the Cretaceous. Arguments about the source of that origin and its subsequent evolutionary pathways are also raging between different specialists. They are all very lively topics.

The five curves show the first sign of a break in the monotonous vegetation cover that most land surfaces had supported since the beginning of the Jurassic 200 million years ago. It was dominated by the thick-skinned trees of cycads and conifers, *Ginkgo*, and ferns. Mostly it was a tropical climate and landscapes were unchanged through millions of years. The C–T eruptions were a major threat to their survival and mark a radical change in the way life was ordered. The onset of flowering plants as major components of the forests brought on more complex ecosystems as evolutionary rates increased for many groups of organisms.

There is another reason for my excitement on the train near Westbury. It was the first time I had cause to feel that my research group's new approach might succeed. Who knows what patterns might emerge? What might we find in other large databases that I knew were being built? If this first set of results gives a significant surprise, what might we find at other moments in time, particularly at the boundaries between two periods? Are there statistically significant patterns lurking within the huge spreadsheets of data? What if the data show up groups of names with common attributes? Perhaps there are clues about evolution and taxonomy. It's moments like this that make science one of the most satisfying things I can think of doing. It must be great to win a big race, score the winning goal, give a fine performance at a concert, cure a really ill patient. For me, the kicks come from having crazy ideas that may come together and make sense.

The realisation of a big expansion in occurrences of these plants 90 million years ago fits evidence from other sources. Around the Pacific rim there were a number of thin patches in the Earth's crust covering deep hot spots waiting to blow up from inside. Most of the volcanoes all erupted at around the same time at the end of the Cenomanian 90 million years ago, and this drastically reduced the oxygen levels from the world's oceans. It also led to high sulphur dioxide in the atmosphere, global warming and acid rain. This sequence of events killed the dominant trees of those forests, the conifers. Sure enough, the curves from our database show a huge drop in the records of pine just before the five flowering plant groups increased. The deforestation caused by the volcanic action was the chance for which the angiosperms had been prepared, with their much more sophisticated ecological tolerance and stronger reproductive abilities.

The five plant groups had originated many million years earlier. The genetic recombinations, the new biochemical pathways that they followed and the first physiological adaptations to the environment all happened long before. These early traits were becoming tried and tested on a very small scale. Modern inventors do the same with their prototype models, making sure the thing works in all circumstances and making adjustments when things do go wrong. Then, when the time is right, a full-scale sales campaign launches the new product into a new gap in the market. It was like this at the C–T boundary 90 million years ago. The five prototype groups of new angiosperms had been tried and tested for millions of years and they worked well, though opportunities were limited. Suddenly, lots of new space opened up where once there had been conifer forest. The explosion of numbers of individual angiosperms had begun.

When I first noticed this sudden fall in the number of records of pine, another thought about the potential value of database mining came into my head. It was an application of my inversion of Lyell's principle of uniformitarianism, mentioned in chapter 1. If the fall in the occurrence of pine has such significance in the Cretaceous, could the present-day falls in occurrence of so many species have comparable consequences? If falls in occurrence in the fossil record show clear trends, do the same trends show up with Red list species today?

Meanwhile, there's more excitement back on the train from Westbury, though by now I was just coming into Exeter. It was not until late in that same journey that I happened to glance at some of the other curves in my collection. Just at the beginning of the Tertiary, 65 million years ago, several of my curves showed sudden increases in the occurrence of flowering plant Families. Most of the temperate angiosperm groups which are known as the Arcto-Tertiary elements showed a clear response to the change: there was a massive increase in their diversity.

The K–T catastrophe

For the dinosaurs, ammonites, and lots of other groups, life ended after an event that happened in Mexico one warm spring day. We know it was a spring day because a magnolia flower was found in the wreckage. A chunk of rock from outer space, 20km in diameter, had hit the earth off the Yucatán peninsula in south-east Mexico. If the meteorite had splashed in deep sea perhaps it would have been different, but the shallow-water impact made it one of the most severe physical crises in Earth's history. The explosion shook the planet and the firestorm quickly spread up across most of North America. The fall-out of dust and smoke was blown east to Europe and beyond. We know that the projectile itself had come from a westerly direction because bits that broke off during the descent through the atmosphere have been found in sediments of the Pacific Ocean.

Flames and heat fanned around the planet. Rock and soil, steam and vapour, charred splinters, roast meat, were scattered into the acrid atmosphere right around the northern hemisphere and most of the south. Weather patterns changed, there was no light on the surface of the Earth. Life stopped or went into hiding; nothing thrived. The thick clouds of debris took many years to clear; the burnt vegetation on land, the acid and ash raining into the sea, halted the majority of life.

There are no graves like Pompeii's, no swamps for easy preservation, so most of the remnants are gone and we are left to speculate. We don't really know how long the fires lasted, their power, their full geographical extent. Like accident investigators or forensic scientists, paleontologists sift through the wreckage of 65 million years ago, looking for evidence. But in this case, we found the evidence before we understood the cause. To make matters worse there were very few refugia, places to hide, that survived the inferno, and of those many have broken up through the intervening 65 million years. Many are yet to be discovered as the fossil record unfolds.

After the explosion the most vulnerable passengers on Earth were the largest animals, the ammonites and dinosaurs, and none survived. The oceans lost much of their dissolved oxygen, so most marine species died within weeks of the impact, and became extinct. Once the black clouds and the fires had subsided new life came to the planet. Some of the carnage itself, both on the land and in the sea, remains as sediment for us to explore and thereby to understand more of what went on through those terrible times. It is at the Cretaceous–Tertiary boundary 65 million years ago, commonly known as the K–T boundary (the K coming from the German 'Kreide', meaning chalk), that we see remains of the catastrophe preserved in rock outcrops from many parts of the northern hemisphere, especially north of Mexico, to the east of the Rocky Mountains.

The thin band of sediments is rich in iridium, an element rare on Earth but common in some asteroids. There are glass globules, fractions of a millimetre in diameter, the relics of molten silicates after the explosion. It also contains burnt vegetation, as well as the rare spring flower. We can even detect remnants from the lump of rock itself, several kilometres beneath the surface of the Chicxulub crater off Yucatán.

It's only in the last few years that pretty incontrovertible evidence for all this has come together. Scientists from very different disciplines have become involved, from all over the world, bashing the story into shape. The idea of the K–T impact event began in 1977 from a coincidental conversation between two people from different disciplines. Walter was a young field geologist and he was talking about a new specimen with a famous Nobel physicist called Luis. It was strange that it should happen at all: geologists don't usually talk to physicists. But in this case they were father and son, the Alvarezes. So the contact was accidental, not part of a planned experiment.

Walter Alvarez had collected a chunk of rock from near the mountains of Umbria, north of Rome. It had three layers: a chunk of white limestone, a thin layer of clay, and then more limestone, red this time with no fossils. The lower white layer was full of microscopic seashells well known to be Late Cretaceous in age. The red rock was like that found all over the region, known to be Early Tertiary. The clay lay between these two differently dated rocks. Did it come from the Cretaceous or the Tertiary? Were all three layers laid down in one continuous sequence? Or does the clay layer represent some kind of break in the process, a gap?

Walter knew that his father's laboratory was testing a new method of dating rocks that contain traces of heavy metal elements. The rock specimen is bombarded with neutrons, breaking down the metal's atoms and making it radioactive. The level of radiation gives the rock's age. The Alvarezes' intuition inspired them to test the Umbria sample for metal elements. Iridium is rare on this planet, only occurring in very small quantities, so it came as a great surprise when the clay layer showed relatively massive quantities of the metal, nine parts per billion. Suddenly the dating experiment didn't seem to matter. The important thing now was to explain the high concentration of iridium. What made that observation really exciting was the knowledge that most iridium comes from space, carried on Earth in micrometeorite showers. So this was obviously some shower. Even greater excitement lay ahead when they had the fragment dated by the new method. Luis Alvarez's rock sample from Umbria was 65 million years old, from the K–T boundary itself.

From calculations based on the amount of iridium in the sample, the Alvarezes' colleagues in the University of California at Berkeley proposed a theory that a 20km-diameter extraterrestrial object hit the Earth, spreading the iridium it contained in its spray. Throughout the 1980s and 1990s what we think is the same 'iridium layer' has turned up in North America, Essex, Denmark, Asia and elsewhere. A speculative idea came from a small amount of evidence and some highly imaginative but testable ideas. What began as a risky expression of grey fantasy by the Alvarezes has turned into a largely black or white fact well supported by other evidence from a variety of greyer disciplines.

Not least impressive is evidence that buried remains of the meteorite have been detected in Mexico. Rediscovered is a better word, for the structure was first mapped in 1962. Then, the ash and lava flows were thought to be part of a normal volcanic structure. There was no reason to get excited and the maps were filed away for normal everyday use. Then, in 1981, five years after the Alvarezes' suggestion of an impact, two petroleum geologists found the Chicxulub crater off the Yucatán peninsula. Impact craters don't lead to oil, so they were told to go away and get on with something else, not quite in disgrace, but not knowing how close to fame they were.

Other evidence of the impact includes glass beads that have also been found in the thin iridium layer sediments throughout the world. It's been known for centuries that when you shoot a hot cannon ball into sand the energy melts the sand and the glass forms into small beads that splash out all around. With big chunks of rock like the Chicxulub meteorite it happens on a different scale, producing a wide variety of different minerals. Some of them form crystals less than a millimetre in diameter and are made of very particular materials such as magnetites containing nickel, as well as quartz, only found in remnants of meteoritic material. A range of clues like these can be identified very easily by looking at the shape of crystals in a scanning electron microscope. Very unusual crystals are well known in the scientific literature and are found frequently around meteorite impact craters of very different ages. They've also been found on the moon, where of course there's no shortage of such craters. Other larger beads are found in the iridium layer at several localities as well as at the site of the impact, while others are pure glass.

They say that out of the space race came the Teflon coating for frying pans. Out of the Cold War came evidence that the K–T impact represented a force much greater than the world's nuclear arsenals could have offered. The energy from the collision reacted on another mineral that's very common on this planet, quartz, which is found in very many types of rock. When they've experienced shock from a high-energy event small grains of quartz show characteristic scratches on their surface. These have been found on grains besides meteor craters and have been simulated with really high-energy blasts in the laboratory.

There appears to have been only one other source that can generate the energy that's required, and that's the nuclear bomb. Sure enough, the same scratches show up from nuclear test sites. You get different kinds of scratches from less energetic explosions and you need a very powerful microscope to tell the difference. The high-power scratches have been found only in Central and North America. Quartz grains with lower-power scratches are common elsewhere from less energetic events.

So, what of the biological evidence for the K–T meteorite falling at Chicxulub? How does that help put a date on when it happened and reconstruct something of the environment, its fauna and flora? There is a lot of evidence, and hundreds of scientists have been working at it and writing about it for the last couple of decades since the Alvarezes' announcement about the Italian specimen. There is much more evidence than from the sudden dinosaur and ammonite extinctions: a panoply of biological change was set into action. There is even some familiarity with the succession that took place after the forest fire from the 1980 eruption of Mount Saint Helens in Washington State.

Both groups, dinosaurs and ammonites, had been showing signs of becoming less diverse for millions of years before the K–T catastrophe. Figure 2.3 shows the number of their Families changing through time. Dinosaur diversity peaked three times, 200, 150 and 80 million years ago – three surges in diversity, three falls in Family numbers, three changes in environment. Changes in the marine realm were different, and the ammonites peaked according to that different tune. If any of these large groups of organisms reach a steady state in a stable environment, there is no need for evolutionary change. But through this same time interval chemical changes in the genes may be going on inside the cells, without expression outside.

If however there's a mishap in the environment caused by anything from bad storms from tectonic activity to a meteorite impact then

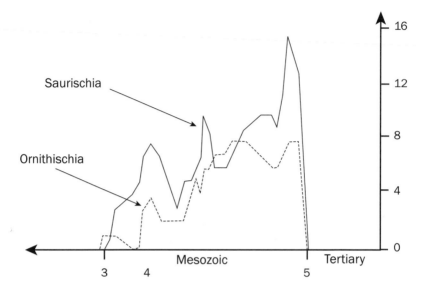

Fig. 2.3 The number of Families known at every one million year interval through the Mesozoic for the two main groups of dinosaurs, the Saurischia and Ornithischia. The last three of the Big Five mass extinction events are numbered on the horizontal axis. (original compilation)

whole groups of different sizes become extinct in different places. This encourages other life, so the diversity curve rises, and dinosaurs recovered from the mishaps throughout the Jurassic and Cretaceous. The final catastrophe however was too great.

Images of dinosaurs and ammonites were well known partly because of their impressive size. Their sheer scale is attention-grabbing. Their size may be one of the reasons why they were becoming extinct just before the end of the Cretaceous period, even before the final catastrophe. Other smaller, less well-known animals became extinct as well. After the K–T catastrophe, many species of microscopic plankton suffered more slowly, and the world's oceans gradually accumulated ash and lost oxygen.

Many other micro-organisms had resting stages in their life cycle which helped resist the hostile conditions: those that didn't died. As well as ammonites, another group to lose many species were belemnites, bullet-shaped animals living in shallow sea. Their ecosystem was stirred up by the rapid increase in temperature and they had no resistance to the changes in environment. Bivalves survived, as did most other invertebrates, the losses being at the species level, only to be replaced by others.

One major change is clear: large land animals were gone. It was to be many millions of years before mammals would get bigger than small dogs. The absence of large animals is one of the greatest mysteries of the fossil record through the first half of the Tertiary period.

First recoveries

Most dinosaurs ate either plants or animals; few were omnivores, devouring both. Herbivores with small mouths selected particular plant species, while other species ripped up everything they could swallow. Some time before the C–T boundary, 90 million years ago, that meant conifers and ferns. Subsequently the tastier, more tender angiosperm leaves may have encouraged these Ornithischia forms, small to very large bipedal vegetarians, for we know they diversified then (see figure 2.3). The other major group of dinosaurs, the Saurischia, contained the bipedal theropods, the carnivores, preying mostly on the other dinosaurs.

But suddenly, 65 million years ago, the dinosaurs were gone, both the carnivores and the herbivores. Although most plants were burnt to the ground by the fireballs that followed the impact, and although the air was dark and smoky, halting photosynthesis, their roots survived. The environment responded to the crisis and quickly recovered. No longer were the conifers and ferns harvested by these hungry foes, the soil was the richer for the forest debris and its microbiology boomed. The temperature of the atmosphere increased and it started to rain very hard in places where it had been drier. The changing environments encouraged the new flowering plants to evolve very quickly.

With warm productive ecosystems on land, in the marine realm phytoplankton were major benefactors from these big environmental changes. Microscopic organisms in the sea, soil and air are especially able to adjust to changes very quickly. Small organisms have a much simpler structure and physiology, more vulnerable to most changes, yet more able to recover quickly. Without oxygen most species became extinct at the C–T and K–T events, but those that didn't quickly recovered and the empty space helped those species to evolve very quickly. There is a sharp delineation at the boundary where some became extinct and others originated in their place. The algae continued to photosynthesise, gathering energy from the sunlight and converting it into food and oxygen, eating up carbon dioxide in the process, clearly a very important stabilising role in the planet's environment. They had done this through the Cretaceous and before, so we know a lot about the great diversity of the microscopic creatures.

Most small mammals also survived, hiding from the heat, being protected by their own sense of exploration. Within a few years some of the planet's ecosystems were beginning to host a new range of animals, plants and bugs. Life began to assume a new normality. Most important of all, there was not a serious loss of the range of DNA, so many branches of the tree of life were able to continue and recover from the cull.

Out of adversity there is usually opportunity, and there was a really creative aspect of the catastrophe. Those organisms that did survive were able to find new opportunities to express structural adaptations. They were able to evolve through the mixing of genes or their mutations that had been taking place quietly through the millions of years before the cull and immediately afterwards. Because the environment had changed very little before the catastrophe there had been no opportunities for these molecular characteristics to express themselves. Evolution was going on inside the cells, in the genes' DNA, and was not showing up in structural features like the colour of a mammal's eyes or a flower's petals. It was as though a strong genetic metal spring had been winding up, collecting energy for millions of years, and then at an instant was released. It caused quick increases in the species diversity of those animal and plant groups that had been inhibited in the wrong environment with its attendant dominant groups of competitors.

Something like this was recognised by Darwin himself, unaware as he was of genes and DNA. He called it 'preadaptation'. Stephen Jay Gould, usually very good with words, called it 'exaptation'. The process is at the centre of the adaptive evolutionary mechanisms, and works within the limits of the fitness landscapes, enabling biology to respond to environmental changes and evolve. Could it be that just as the environment appears to have changed in sudden bursts, separated by millions of years of quiet calm, so organisms respond with matching steps of structural change, either extinction or radiation, and stasis?

It appears that mass-extinction events happen at different times for different reasons and with very different severity and effect. We know that each event is different and none can be predicted; nevertheless they do have things in common. The events are triggered by environmental changes, possibly from fire and flood, so reducing light and oxygen to slow down photosynthesis and respiration on land and in the sea. The consequent culls usually lead to vacant ecological riches which are eventually occupied by new forms that have adapted to the fresh conditions.

Perception of this intimate relationship between environment and biology is much stronger than it was even a few years ago, caused by bringing together ideas on biodiversity from previously isolated subjects: geography, geology, biology, physics and chemistry to name but five. Evidence from these disciplines helps us see events that force environmental change, and which then become the principal causes of extinctions. In turn they offer new opportunities for the newly stored biochemical and genetic developments to spring into action and create new species within the new ecosystems. Changes in these cellular processes are the eventual response to environmental attacks from things like asteroids, volcanic output, sea-level change, atmosphere change, and oceans with little dissolved oxygen.

Looking for trends

It was especially in America that paleontologists were able to accumulate evidence for regular catastrophes by looking at the plentiful outcrops of sediment east of the Rocky Mountains. The popularity of the stories of the dinosaurs' decline and the mammals' rise added to the public interest and the money available for more studies began to rise through the late 1980s and the early 90s. Lateral-thinking grant-seekers were even funded by NASA on the ground that the aftermath of the Cretaceous–Tertiary (K–T) catastrophe might yield clues to understanding the aftermath of a nuclear winter arising out of Ronald Reagan's Star Wars policy.

This was because a little earlier, in 1984, the debate took a turn in an unexpected direction. In February of that year David Raup and Jack Sepkoski, prominent paleontologists at Chicago, published evidence that suggested to them that something like the K–T meteorite hit the Earth every 26 million years. They had a large database of the times of the first and last appearances of marine invertebrates in the fossil record. Their graph shows steep extinction peaks 65, 39, and 13 million years ago, and then as far back in time as between the Permian and Triassic 250 million years ago. Their bold suggestion was one of those risks that scientists rarely take; others sat up and noticed. It certainly directed a lot of thought and energy on to the deep relationships between environmental crises and evolutionary changes.

Not least lateral of the thoughts were those from the American astronomy community who felt challenged to react to the 26-million-year-cycle theory. Might there be a regular visitor past our planet or solar system with this same frequency? The massive computers began to crunch their data. Then the data were thought to be big and the computers fast: now we know otherwise.

Just two months after Raup and Sepkoski's announcement in the *Proceedings of the National Academy of Sciences*, *Nature* magazine published five articles by ten American astrophysicists. They were responses to the February paper, published unusually quickly and so attracting a lot

48

of comment about their relevance: how could they have been refereed so quickly? It started a row about how results and publications are validated, and the argument continues to rage, especially because future funding is influenced by the process of peer review. Two of the April 1984 articles argued for a dwarf companion for the sun, two favoured oscillation of the solar system up and down the galaxy, all these more or less happy with the 26-million-year time span. The fifth manuscript favoured 28-million-year-cycles from evidence of the age of craters. Since then we have not heard much from astrophysics about the K–T event.

The response to Raup and Sepkoski's theory from outside America was much more cautious. Special meetings were held in Europe, Australia and Asia to summarise the state of knowledge about extinctions for the major groups of animals and plants. There was a strong atmosphere of excitement at the UK Systematics Association in Durham in September 1986, which was to share the evidence for extinction patterns from specialists in different groups of organisms. The gathering got off to a good start with a demonstration of how dinosaurs mated. Bev Halstead was an eccentric specialist in fossil vertebrates. He and his wife Ann had worked out the movements by studying footprints in an Oxfordshire quarry, so their mimicry wasn't entirely imaginary. They had their audience in stitches, Bev bouncing on top of his not-so-cooperative wife, roaring in agony when her tail got in his way. The sharp plates and long spines didn't help him either.

At the beginning of the meeting there was a buzz of excitement at the idea of some influence from outer space on the process of evolution. Expectations were very high because few had ever considered looking for extinctions, the emphasis always concentrating on the first appearances, the origins. The excitement even beat the Halsteads' demonstration of dinosaurs copulating. Evolutionary biologists had tended not to devote much attention to the negative effects of environmental change, influenced no doubt by the title of Darwin's classic. Perhaps the optimism of human nature puts origins before extinctions. Nevertheless, however frequently they occur, and however poor the fossil record,

49

the many different mass extinctions were beginning to follow common trends. We had met in Durham to see if we had much evidence in Europe to support the theory of 26-million-year cycles.

The major groups of marine and terrestrial organisms were each represented by a leading specialist who had been asked to look for trends in the timing of extinctions. We left the meeting, however, with ideas of quite different changes. A lot of the extinctions seemed to happen at the same times, and there was more talk than usual of environmental changes being associated with these extinctions. Then after these same environmental changes there seemed to be other species following a process of recovery. In 1986 there seemed to be no rhythm or reason to it, but we did seem to be edging towards an idea of some organisation or pattern.

In those days the paleontological community was divided about the theory of 26-million-year cycles. There continues to be a strong reaction to the flippant use of fossil data, with insistence on the use of sophisticated mathematical and statistical validation to quantify quality. There is also great interest about the unreliability of the fossil record and distrust of observations that some marine species disappear through an environmental crisis only to reappear when the crisis is over. These are called Lazarus taxa, returning from the dead. The commonest Lazarus taxa are molluscs and corals. Both these types were very sensitive to environmental change and so they respond very clearly to extinction events. Those individuals that survive appear to fall to very low numbers and show up again only in refugia protected from the environmental excesses. Perhaps their small numbers are too few to be found as fossils. Eventually the environment recovers to some kind of normality and all available ecological niches are occupied, though not necessarily in the same way as before. It's what happened to Lazarus while he was dead that is a question open to speculation.

Then, in 1993, the same two American scientists made another more popular suggestion about the end and beginning of major animal groups, referred to as the Big Five extinction events (see figure 2.4). These came from Jack Sepkoski's data, based only on the marine realm.

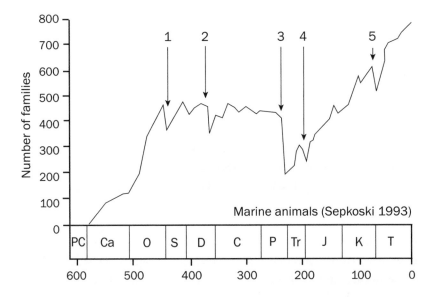

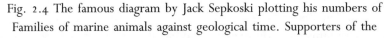

Fig. 2.4 The famous diagram by Jack Sepkoski plotting his numbers of
Families of marine animals against geological time. Supporters of the
'punctuated equilibrium' model of evolution see the Big Five
mass-extinction events separating the relatively quieter phases of evolution.
Supporters of the exponential model see an exponential curve pass through
these data points, as drawn in figure 3.5.

They gave five particularly large extinction peaks, though there are
plenty of others. Sepkoski puts them at 439, 367, 245, 208 and 65
million years ago. None are within a 26-unit rhythm but they are all
close to major boundaries of the standard stratigraphic scale. These
had been suggested in the early years of geology by Lyell and his
contemporaries, following the canal surveyor, William Smith, largely
from breaks in the rock types that show up on cliff sections. Our much
more sophisticated data appear to be confirming the early statements
with patterns from changes in environment and biology.

In America there is more interest in science from the public, and
pressure exists to expect the *biggest* and the most dramatic, the *longest*

and even the *oldest*. So the Permian Triassic (P–Tr) boundary is seen as the *biggest* extinction event. The Big Five is a good soundbite but their selection and the preferential sequence depends on how they are measured, what they comprise, and how they are caused. Mass-extinction events were not alike in cause, duration or effect and really defeat this kind of treatment. Still, the Big Five sticks because it sounds good.

Another consequence of the snappy soundbite is that it tends to make people believe it's true. Recently politicians, journalists and the public have expected scientists to know everything about mad cow disease. But science is not like that. We cannot know, we can only create an intelligent guess and then try to prove the guess wrong. So many things that change through time show different kinds of fluctuation in different ways. In modern life, economics, fashion, the stock market, all show ups and downs, in many indices, through different intervals and limits. For each variable there are many ways of monitoring the changes. The fluctuations depend on what it is you measure, just as in natural systems, like the weather. Wind direction, atmospheric pressure, temperature and so on will give different-shaped changes. Some may be related, some may not.

Through geological time the main internal character forcing these environmental changes has acted on the planet's surface. There are active movements of plates that form the Earth's crust (see figure 2.5). These plates are on the move, geographically and in depth, driven by convection currents from the much hotter and more fluid inner mantle on which they float. Additionally, molten rock from beneath the Earth's crust pierces through the skin like pimples, while the plates are much wider in extent. Most of these igneous spots come from under one region like Iceland, or the Deccan in north-west India. Plates cover large areas, pushing rock up on one side such as the mid-Atlantic ridge, and down into the mantle on the other side. The movement is usually no more than just a few centimetres a year. Through long periods of time these disturbances affect ocean current, which affects weather systems and so on. The reactions interact but not necessarily like a

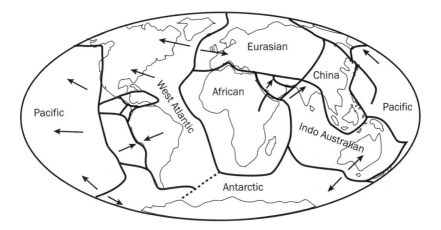

Fig. 2.5 A simplified map of the Earth's tectonic plates. (after D. H. and
M. P. Tarling, *Continental Drift*, London: Bell, 1971)

chain. Sometimes an ocean may be cut off from another, dammed by
the meeting of two plates, to cause all manner of changes. At other
times extra rock might be pushed up onto the ocean floor, raising the
level and pushing the water higher.

For example, about 10 million years ago the moving continents of
Africa and Europe pushed Morocco into Spain, rather as India had
crashed into Asia earlier. The Strait of Gibraltar closed shut. You could
have walked from Glasgow to Cape Town. At the same time there
were fluctuations in the concentration of CO_2 in the atmosphere and
the resulting greenhouse effect caused the water in the landlocked
Mediterranean to evaporate. That's why there's so much salt in the
region, and the Black, Caspian and Aral Seas are relics from the same
complex changes. Plate tectonics can easily have a major impact on
the environment and can cause major changes very slowly and very
quickly. One moment a sea, the next a salty desert. And when the
plates drifted Spain away from Morocco again, a few million years
later, imagine the giant waterfall at the Strait causing another big change
for the plants and animals. Environments, climates and weather can be

changed very easily, and most changes happen through different kinds of processes and combinations of processes. Every mass-extinction event, let alone every mini extinction event or protracted change, will have been due to a unique combination of processes.

The mid-Atlantic ridge spills molten rock from the Earth's deep inner mantle onto the ocean floor, warming the sea and creating its own peculiar deep sea ecosystem. When these outpourings at the edge of continental plates are on land, a different kind of havoc is let loose, volcanoes. As well as changes to the substrate for plants and animals, to drainage and land forms, the atmosphere especially comes in for a real shock. There is ash and smoke, gases like CO_2 and SO_2, and burnt forests and soils. The lava is hard and hostile, no place for these products to sediment, so they run into the sea. In these rare circumstances the land has been transformed directly by the lava on the surface and the thick clouds in the sky stop photosynthesis. Life there is threatened severely and the marine organisms are also killed off by the long darkness, high acidity, the black organic ashes and the resulting lack of dissolved oxygen. Without bugs in the soil, and plankton in the sea, the larger forms of life can't keep going. The cycles stop.

Today's knowledge of the fossil record shows that the largest number of extinctions took place about 250 million years ago at the P–Tr boundary. The latest estimates of species becoming extinct around then give figures to 90 per cent for all species of marine animals, 70 per cent for terrestrial vertebrates, and more than 90 per cent for land plants. Explanations abound, even more ebulliently than for the K–T boundary. For the last decade, most geologists have had the view that this biggest-ever loss in species took place over a period of 11 million years. Volcanoes erupted as never before or since, giving lower temperatures at first and then a greenhouse effect from the high concentration of carbon dioxide. Huge flood basalts dating from this time in Siberia formed the largest flows of basalt in the world. There is no doubt about the scale of the catastrophe, so the question is not 'whether' but 'how?'

Early in 2001 a new breakthrough in understanding this P–Tr event was published, suggesting that it was quick and spontaneous rather than lasting several million years. Whiffs of the original atmosphere from the time of the event have been found trapped in small bubbles from rocks in China, Japan and Hungary. At each site they found a particular proportion of the two rare gases helium and argon, a proportion known only in meteorites. But P–Tr boundary specialists are yet to find the crucial supporting evidence we have learnt to expect from mass-extinction events. Where is the impact site and its buried bolide?

However it was started, the resulting global warming caused further threats to land and sea organisms. In particular, near-shore habitats suffered from the resulting change in sea level, which also meant that much less dissolved oxygen. The landmasses were also exposed to these forces, accounting for such large numbers of terrestrial animals and plants becoming extinct at the same event. The end of the Triassic period was marked by another sudden change in sea level and a consequent extinction of marine animals. Then you find yourself on the steps at Lyme Regis pier getting onto the rowing boat, where this chapter started.

After these stories of the P–Tr, C–T and K–T events I give you a hypothesis. Now we have another mass-extinction catastrophe, caused by one species: us. Many of the signs are familiar to geologists – changes in sea level threatening to change low-lying coasts, changes in climate due to greenhouse gases in the atmosphere, the fall in the water level in enclosed lakes. The glacial-interglacial cycles of our present climate system mean that we are naturally alternating from icehouse to greenhouse very quickly, every few thousand years. More threatening still is the effect of humans on this changing world. In February 2001 the Intergovernmental Panel on Climate Change issued its third assessment report. It says that environments are going to change much more severely and quickly than was thought. We are living through a major environmental crisis, something looking very much like a mass-extinction event. But there is one trouble with this hypothesis: where are the extinctions?

3

A System out of Chaos

White noise in the universe

It's hard to believe that all the evolutionary and environmental changes at the K–T boundary outlined in the last chapter were started by one random event. If the 20km diameter meteorite had not hit the Earth 65 million years ago none of it would have happened, because the forces of environmental change and evolution would have gone in a different direction. The even more cataclysmic earlier impact 245 million years ago, at the P–Tr boundary, now appears to have been caused by a similar but much larger object, causing the biggest upset that life on this planet has ever experienced. But what went on between these large events, let alone before the Earth became so complex?

There have been many smaller upsets to the environment as well as to biodiversity, some more regional than global, and most stimulated by random hits by outside objects or by internal events like volcanoes erupting. Between these crises there were much quieter and more regular periods of small environmental changes caused by slowly-moving continents and steady changes in climate and sea level. At times like these there are relatively few evolutionary origins or extinctions. It has been more usual for the planet to pass through long times of quiet when nothing much happens. But there are contrasts.

The most extreme example of this simple state came immediately after the Big Bang, when the universe was a very strange place. Space and time were created and matter was infinitely hot. It was about 15,000 million years ago, when the universe was a second old. Protons, neutrinos, photons, neutrons, electrons and positrons exploded outwards and cooled down to 10,000 million degrees. It was too early for atoms to be stable and many changes were taking place. A hundred seconds later, protons and neutrons started to combine, each pair giving a nucleus of deuterium, or heavy hydrogen. This was a process that was to be repeated many times, then giving hydrogen, helium, lithium and beryllium. Nuclear scientists now calculate that the production of these elements stopped within a few hours of the Big Bang. The astrophysicists say that for the next million years or so these first elements of the universe just kept cooling and expanding outwards. Nothing much else happened.

Eventually the universe began to produce galaxies, and the process was to create the recognisable structures of the universe we are beginning to understand today. It appears to have form, but not predictability, and physicists are formulating new ideas about the way the process happened. This involves mathematical analysis of data from astronomy and difficult predictions from theoretical ideas.

One crucial concept in these thoughts is 'chaos', a word with two meanings. In the traditional sense it is disorder, just a constant randomness hanging on as the universe cooled and expanded. The system was like the hiss of the radio signal once a station has closed down. A plot of the frequencies shows no bumps or patterns, just a continuous boring squiggle (see figure 3.1). Another name for this signal is white noise. It is chaotic, unpredictable and extremely sensitive to the initial conditions of the system. Throw ink into a fast-flowing stream and its chaotic spread will quickly be seen. The early flow determines the direction, and eddies may disturb the forward movements. There is no sense or design to the resulting mixture.

The second meaning is more apposite to our understanding of how things continue to develop from this disorganised state of affairs. It describes a developing sense of complexity, resistant to analysis and prediction. The theory can be applied to a lot of geological data, and the fossil record is particularly amenable. It suggests that through time increasing diversity of species leads to the increasing complexity of the

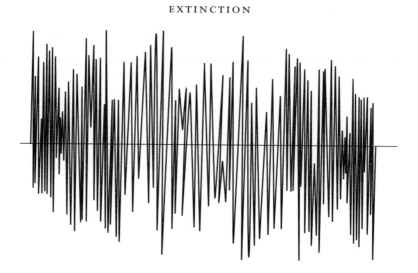

Fig. 3.1 A signal from white noise (after Bak, 1996). Different frequencies signal through time, with no clear patterns emerging. It is like the hiss of an empty radio channel.

Earth-life system. And with our increasing knowledge of the fossil record and ready access to computing time we can begin to create algorithms that can break these rules and make predictions.

This transition, from the white noise of a boring squiggle near the beginning of time to the high complexity of modern life on Earth now, is driven by chaotic processes. They may even become predictable if the system can be in control of itself. It is only influences from outside that will interfere with the forecasts. The process fits with the second law of thermodynamics in which the Earth is passing into a state of maximum entropy with no order at all, the universe at its death. The idea is that at the beginning of time the system had full order, and at the end it will have none.

Now, part-way through this process, we see energy being used to sort out the high order of the past into the messier systems of the future of the universe. It is the transition from white noise to what we might call black noise, and it gets caught up with the process of biological evolution on the changing Earth. When that process is monitored by the analysis of data from the fossil record we see a signal of intermediate frequency between white and black noise. It is called pink noise and is characteristic of self-organised natural systems such as fractals and sand piles, which also give power laws.

These are human concepts which we devise to try and understand how the universe develops, and particularly how life on our planet works. Another concept that has been used is the fractal, comprising the infinite variety of shapes seen in Mandelbrot sets. They are like the fast-expanding duplications of geometrical images you get on some screen-savers when you're not using your computer. At first sight fractal images appear to be highly complex and forever original. But the software algorithms that run them are very short and simple and go round in circles, repeating the same instructions, just one minor difference setting off a whole new direction with the same kind of building block.

One theory is that the early Earth had these simple features of chaos. There were no external forces, no limits of space within the system, no interactions between the particles. The only variables were time, space and temperature, and nothing got in their way. It was a perfect chaotic system in a closed universe. But it had to end because there were other influences. Stephen Hawking has favoured a smooth model for our early universe. As with the millions of other stars, the early simplicity moves to disorder through chaotic changes. This led to the first slow emergence of structures.

Concepts such as chaos and entropy were developed by mathematicians to help physical scientists, not biologists, understand their worlds. It's difficult and dangerous to extend the same thinking to nature, where there are other holistic considerations than purely physical ones. As I will argue later, I doubt that biology is so quantifiable. But because they come from quite different mathematical traditions, to compare chaos with disorder is not comparing like with like. Instead biologists should expect to find their equivalents of chaos and disorder within the large system of life on Earth. It is a timely suggestion, because large sets of data about the evolution of life are becoming available on the internet and soon they will be tested against some of the theoretical models explaining how life can come out of these chaotic systems.

Some parts of the Earth have moved toward greater order as a result of the chaotic changes mentioned here. The complexity of life is counter-entropic, more and more ordered. The present atmosphere with so much free oxygen is counter-entropic. It should all be locked up in chemical compounds and is only kept going by plant life. Chaos seems to enable areas of greater order to be fenced off as life inside the general thermodynamic trend to greater disorder.

Both in the Big Bang and in the formation of Earth, systems that started off by appearing to be random, balls of plasma or dust, were forced to obey the law of disorder. This early state turned out to have organisational properties that produced structure where there appeared not to be any, making suns and galaxies. Physicists have started to explain the Big Bang's process, needing the complexities of quantum mechanics to do it. Earth scientists have pieced together the processes that gave the Earth a core, a hot mantle, continents floating on top of it, and ultimately the beginnings of an oxygen atmosphere, etc. etc. Chaos theory contributes to understanding such hugely complex systems. Life is another such system. Nothing ever goes ABCDE. Every time science gets a handle on one of its processes, some other offers itself, knowledge expands, causes multiply, uncertainty rules. With enough data and with ways to organise the data, patterns can be detected and predictions made. Furthermore, the ability of small shifts to lead to vast changes means that nothing is free from contingency. Disorder rules, and life has managed to bend the rules by sailing on the winds of complexity. Now human beings have become such a force in ourselves. We can alter the planet as no other life form has been able to do.

As well as the meteorites colliding with the Earth to mark the end of the Cretaceous and Permian periods, another much bigger collision occurred earlier. It happened about 4,500 million years ago, just 50 million years after the formation of the Earth and Mars, well before life began. Paradoxically, the evidence for this new idea comes not from the Earth, but from the moon.

The US Apollo moon missions of 1969–72 are rated among the highest of human achievements. The live television coverage of the moon landings and re-launches, the technology, and the emotion of 'one giant step for mankind' rank high in the annals of the last century. In contrast, it's interesting that so little publicity has been given to the results from the scientific studies of the rocks collected. For they are even more exciting than anyone imagined, and offer an explanation of how life came to be here on Earth.

Studies of the rock samples brought back from the moon made through the 1970s and 80s gave a range of opposing interpretations and arguments. Since then, the work has established a theory accepted by most of the participants, that the moon came from the Earth. The theory led on to speculation that the Pacific Ocean is so big because it is where the biggest impact of all happened.

Just after the formation of the solar system another important event happened. Astrophysicists tell us that an asteroid or planet about the size of Mars gave our early planet an oblique glancing blow. The energy released from the impact melted the Earth into layers: atmosphere, mantle and core. The atmosphere comprised water vapour, carbon dioxide, nitrogen, xenon, argon and helium, the core iron and nickel, and the mantle basalt and silicates. The force of the blow on the molten mantle caused a large chunk to splash out. This mass began to orbit the planet, forming our moon. We know it from the rocks brought back by the Apollo crew. Earth rocks of such an extreme age have long since moved back into the Earth's crust down below the tectonic plates, or more simply eroded or weathered into dust. On the static moon they were preserved, changing only in daily temperature.

The moon event determined the rest of the Earth's history. Before the impact the early planet had static climates, from the poles to the Equator, the same throughout the year. Things never changed and there were no seasons. This was because the planet's axis of rotation was truly vertical. The force of the huge impact knocked this to one side, causing the axis to tilt, at present 23.4° from the upright. This angle of obliquity changes through different times, causing oscillations through several thousands and maybe millions of years. The shift in the angle of rotation means that one side of the planet is nearer the sun for half the year, and has summer. The other side has winter, though around the Equator there is little change.

Seasonal changes in climate had begun, and had happened by a random event, one big astronomical object crashing into another. Energy was being worked into this new world. The then living system began to run itself and self-adjust to internal and external changes, just like many other things in nature – your own physiology, for example.

These developments in the universe gathered pace because the expansion outwards created regions of different densities. Gas clouds condensed to give galaxies and further concentrations of gas spun round with increasing force to form stars. They were like clouds in a storm, instead forming stars and their systems in each galaxy. In turn, each of these underwent a period of stability characterised by the physical properties of the galaxies. As the universe and the galaxies and the stars became more complex, the systems showed patterns of random distribution of matter between the explosions after the Big Bang. In system Earth there is a major transition caused by changing environments giving a new kind of order.

Changing environments on a planet with water, atmosphere and carbon compounds can create life and evolution. For these systems to survive, let alone develop, catastrophes become essential features within the complex processes. They initiate progress on the planet from simplicity to complexity and are driven forward by the reactions from inside the system. They have the ability to change the noise from the boring unstructured hiss of white noise to the beauty and orderly complexity of a Bach concerto.

Sand piles of self-organisation

Some of the first ideas about these systems were published in 1987 by Per Bak, a physicist, through the inspiration of a group of scientists in New Mexico. They were announced in the prestigious journal *Physical Review Letters* and quickly became one of the most quoted articles in mainstream physics. But as was the way then with most science, Bak's ideas were heard only within his own specialism.

Bak introduces the idea of self-organisation with a graphic experiment. Fill your hand with sand, all grains the same size and composition. Let it flow from a constant aperture between your thumb and two forefingers. The steady stream of sand falls to form a cone. Everything is equal, the aperture, the size of the grains, weight, surface features, the rate of flow. Through time the cone gets bigger and heavier, each grain becoming part of the increasingly complex system. The cone will adjust to the increase in the number of grains, and react from within. Sometimes a single grain will be holding up a pile of others. Elsewhere, smoother juxtapositions of the grains may create a firmer structure. Inevitably, an avalanche of grains will occur, and it is unpredictable how long it may last, and how powerful it may be.

Such avalanches show a clear mathematical identity and a pattern of change through time that can be plotted as a straight line with a characteristic slope (see figure 3.2). This is called a power law: variables that plot as a straight line from any self-organised system. It is because large changes are rare, small ones are common, and the variation between the extremes is smooth, just as we will see develops on planet Earth.

Bak's article marked the start of something very exciting at the Santa Fe Institute in New Mexico, under the creative leadership of Stuart Kauffman, a medic turned multidisciplinarian complexity scientist. They shared some of the excitement from the early 1990s in their popular books, *How Nature Works* and *At Home in the Universe*, looking for laws of complexity and encouraging specialists from different disciplines to give data, ideas and methods of analysis. They started to dare previously

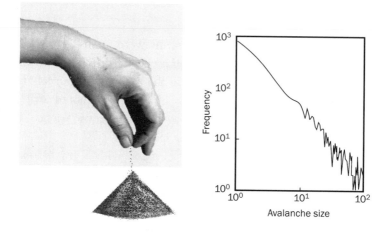

Fig. 3.2 A constant flow of sand forming a sand pile and a plot of the sizes and frequency of the resulting avalanches. There are fewer very large ones and more small ones. The resulting straight line is called a power-law curve. (after Bak, 1996)

outrageous thinking: if liquid water is at the critical point, 0°C, when it becomes ice why should not life also have its own critical point when it hangs between order and chaos? But some force is required to drive the change. That is a major feature of self-organised criticality: being at the boundary between one state and another, and making large to small changes suddenly and unexpectedly. The changes come from forces entirely within the system itself, not from outside. They generate complexity to be absorbed by the whole, as when ice becomes water above 0°C or when high pressure causes graphite to become diamond.

Within the closed system of a sand pile, free of interference from weather and kids playing on the beach, the avalanches happen necessarily and randomly. Measuring their size and timing reveals lots of little sand slides and a very few big ones. If you plot these results as in figure 3.2 you see a straight line expressing the power law. This is a mathematical identity that shows up in the self-organised systems found by Kauffman and Bak. You get the same-shaped curves they found from sand piles in big databases of the fluctuations in financial markets, traffic jams and the notation in Bach's music. A small number of large-scale features are at one end of the straight line and a large number of small-scale features are at the other. In between, the proportion gives a straight change in the inter-relationship. Not only do we find the same sequence in our human-made world, but there are also natural ones like landscapes, earthquake occurrences and weather patterns, which also show this same straight-line identity. It is as though all natural systems behave like this from within.

As well as the sand-pile analogy, Per Bak also uses the traffic jam to illustrate a perfect self-organised system. Unfortunately we are all familiar with roads full of traffic. Often, on a busy motorway, cars have to slow down with one another and accelerate off again. This often happens in waves that are generated from within the system itself. No crash or obstacle from outside the system is necessary, the cars themselves create the changes, by lots of drivers making small hesitations and fewer giving bigger movements. Of course if there is a crash, it shows up in the time-series curve as well, but the system recovers. Once you are hooked on the idea of self-organised systems their features show up in many of the time series you see in nature. How did we miss them earlier?

Gaia

During the early 1990s the broad implications of the concept of self-organised criticality were discussed in detail at the Santa Fe Institute in New Mexico, in groups led by Stuart Kauffman and Per Bak. Now it has been extended to a wide range of phenomena that change through different timescales, and a wide literature has developed the mathematical and more applied implications. Twenty years before, James Lovelock proposed 'that organisms contribute to self-regulating feedback mechanisms that have kept the Earth's surface environment stable and habitable for life'.

Lovelock's next-door neighbour was the Nobel-prize-winning novelist William Golding. After reading *Lord of the Flies*, you might imagine that he found Lovelock's idea quite attractive. 'For a big idea', he told his friend, 'you need a big name. How about naming it after the goddess of the Earth, "Gaia"? It's better than "The cybernetic theory of a homeostatic Earth".'

But whatever it's called the theory needs testing, and Lovelock had no data to test, just a satisfying idea about life on the planet. The realisation that the Gaia theory is untestable upset a lot of reductionists in the scientific community and for a time attracted comments that the main support is for its mystical qualities. However, in 1983 Lovelock regained his scientific credibility by advocating a theoretical model of self-regulation that he called 'Daisyworld'.

It describes a world that has just two kinds of life, black daisies and white daisies. One absorbs the sun's energy and gets hot, the other reflects it and stays cold. The black daisies grow more abundant until they cover all the available land. But the temperature is getting too high for their comfort and so they die off and the white flowers rejuvenate to cool the planet to the optimum temperature. When they thrive in their turn and spread too far, the planet overcools. Then the white daisies start to die off and make room for the black daisies to bloom again and start to reheat the planet. The regulation happens automatically, however the sun's energy may vary throughout the year. So impressed with this argument was one of the world's leading ecologists, William Hamilton, that he presented a mathematical model of his own in which a species could affect its ecosystem, maybe even leading to its extinction. With Hamilton and many others giving their support, the Gaia theory finally achieved respectability in 1998 with a 7-page review article in *Nature* by Tim Lenton.

But the reductionists are still out for blood, or at least, for evidence. There is James Kirchner in particular, a geomorphologist and geophysicist from Berkeley, who likes to measure big things very precisely. But the fossil record and much of modern nature do not provide data comparable in accuracy to that of the human genome project or astronomy. Nevertheless, he persists with his fastidiousness, hoping for biological evidence as good as that from physics. There are many others who use the same reasons to criticise ideas such as Gaia. The same people are unhappy that a power law and pink noise show up from images of self-organisation.

A softer approach is that the data may not be entirely reliable but they can be cause for experiment and empirical debate. I think that Bak and Kauffman's ideas of self-organised systems also have the effect of making a Gaia 'theory' unnecessary. Lovelock's introduction of the idea did a great service to help us see planet Earth from outside. Now we must move on to find more data and test the theory of self-organisation through a more scientific process.

If biological evolution really is a self-organised Earth-life system there are some very important consequences. One is that life on this planet continues despite internal and external setbacks, because it is the system that recovers at the expense of some of its former parts. For example, the end of the dinosaurs enabled mammals to diversify. Otherwise if the exponential rise were to reach infinity, there would not be space or food to sustain life. It would come to a stop. Extinctions are necessary to retain life on this planet.

To investigate that idea let's move into the environment itself to see if we can find evidence that environmental changes have culled whole groups of animals and plants to relict status and even extinction. What better time to start than just as someone kicked the sand pile to cause the huge avalanche of extinctions at the K–T boundary? The story of the recovery is as fascinating as that of the extinctions themselves.

Unmeasured patterns of beauty

These patterns of self-organisation were absent from the early universe because there was nothing to be organised. Randomness gives white noise, complexity gives a clear structure. The first gives a meaningless list of numbers, or a hiss of deafening sound. Whatever the shape may be it is dull and boring, without character and most certainly without structural organisation inside. It may well symbolise some state of affairs decided by others, but it is not leading to any meaning for itself. The second, complexity, is built of different levels in its structure, whatever the materials. The patterns that emerge can become adored as beautiful objects.

The same complexity persists today in the fractals of a Mandelbrot set, flat images playing with a simple pattern, lacking any sense of creativity or interpretation. As soon as some kind of variation or interference enters the system this simplicity is broken. Within our story, it first happened when that huge lump of rock hit Earth 4,500 million years ago. The resulting seasonality caused environments to change. From this an increased propensity to complexity began on Earth. Environmental change such as this was an essential part of evolution.

It is fair therefore to suggest that like sand piles and traffic jams, landscapes and evolutionary biology display power laws in some part of their behaviour pattern. When the frequency of changes in such complex systems is analysed, a different pattern to that from the hiss of white noise appears. It's a pattern with a small number of large changes and a large number of smaller ones: a power law. It is what happens when you watch a sand pile growing – few large avalanches and lots of small ones.

When the actual numbers of changes are plotted against their size we always get a curve shaped like that in figure 3.3, the sign of pink noise. Mathematically this is related to the power law shown in figure 3.2. If the data came from a self-organised system, showing things such as environmental or evolutionary changes, the curve will have this pink-noise shape. It is an expression of the complexity of the system. Just as the simple hiss of an empty radio channel will give white noise, so natural changes appear to give pink noise with the shape of figure 3.3 and a power law of figure 3.2. The first signs coming from recent analyses of data in evolutionary biology show that the apparently chaotic sets of information, when taken together, show these curves representing a power law and pink noise.

68

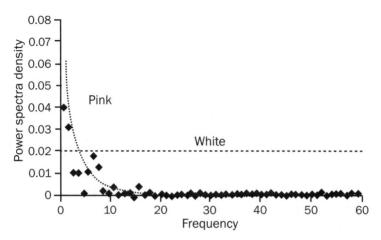

Fig. 3.3 Curve showing the frequency of the time intervals between first and last appearances of taxa in the *Fossil Record 2* database. If the data represented the hiss from an empty radio signal they would give the horizontal straight line of white noise. Instead, because the data come from the natural system of evolutionary biology the curve rises at the lowest frequencies. The shape of the curve is that of pink noise.

These are new concepts, and we have to be very careful about what these patterns actually mean, let alone how they relate to one another. We must approach these different worlds of complexity and chaos very gently until our understanding has been tested from many more large sets of data. Is it the confrontation between subjectivity and objectivity; between beauty and plain geometry? After all, Plato suggested that there were connections.

Measuring patterns of evolution

Back in the early nineteenth century, a few thousand miles south of Stuart Kauffman's and Per Bak's Santa Fe Institute, Charles Darwin was having revolutionary thoughts about the unique species surviving on the Galapagos Islands. He spent a month there in 1835 making the first observations of the many relict species of plants, birds and large animals. Alone in his cabin on the HMS *Beagle* he started to think about how the strange creatures found there had reached such remote islands, and why each island had such subtly different inhabitants. Two years later, when Darwin was twenty-eight years old, he was arguing publicly with his mentor Charles Lyell about why the Galapagos finches were so slightly different on each island. Were they different before they inhabited the islands, or did the different beaks form once they had arrived?

The question needs the differences to be measured and compared in detail. Darwin saw evolutionary change as going on evenly, at a smooth and constant rate. The single illustration on his *Origin of Species* is an evolutionary tree (see figure 3.4) that's symmetrical at all its branches. It follows that he thought that evolution was steady and gradual and that the rate of evolution followed a straight line. In those days, there was so little known that it was the only possible explanation for evolution to proceed, in this linear way, with continuous adaptation, selection and mutation. The discoveries of changing ecosystems are about as new as the discoveries of changing nucleotide base sequences, and they don't happen evenly or equally.

Without knowledge of genes or DNA, let alone ecology and the distinctive features of island biogeography, evolution was a hard case for Darwin to argue. Lyell's developing ideas about the scale of geological time helped a little, but neither man was confident about the answer. The pressure from the social establishment against their embryonic ideas was immense. There was so much ignorance about what was needed to stimulate their thoughts. At least they began to realise that the major units of geological time had differing lengths, but not that events happened at different frequencies. Why should they have realised that evolutionary change might not happen at an even rate? There were to be more than twenty years of arduous observations and thinking about more pertinent issues before Darwin would find enough evidence and courage to publish the first edition of the *Origin*.

Through more than a century, biologists hung on to this same implicit

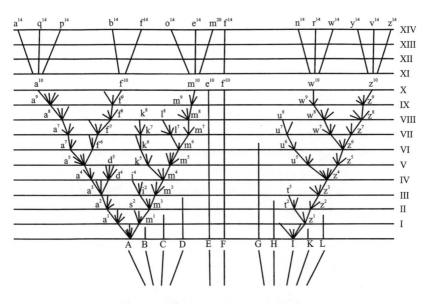

Fig. 3.4 Copy of the only figure in Charles Darwin's 1859 *The Origin of Species*. The two evolutionary trees show branching at each interval of time. A and I are the first species in each tree, which diversify to different levels in the lineages represented by each lower-case letter. The time intervals on the vertical axis are all the same size, suggesting that Darwin thought evolution took place at a linear rate. Compare this diagram to figure 5.1 from Louis Agassiz's 1833 *Recherches sur les poissons fossiles*.

assumption that evolution occurred evenly. Most biologists didn't really speak to other scientists such as those we now call environmentalists, ecologists and geologists, so the effects of one discipline on another were not appreciated. Any suggestion that genetics might have something to do with environmental change was not thought to be important and was rarely considered. The great biologists of the early twentieth century didn't have any evidence to put against Darwin's original assumption of a constant rate of evolution. But it was a non-issue because no one gave the question of evolutionary rate any attention. What did happen was that a number of new species were being added to the fossil record. It was seen as a simple linear system that depends on the number of species at a particular time. From their writing you can see that most twentieth-century biologists tended not to look outside their own specialism, certainly not to look at the life system as a whole.

Since the middle of the last century scientific advances have continued to occur within single disciplines. For example, in chemistry Watson and Crick found the structure of DNA that enabled the genetic code to be understood and sequencing to begin. The ornithologist Robert MacArthur began quantitative ecology. Meanwhile evolutionary processes remained enigmatic. People still talked in Victorian aphorisms, as some still do, such as 'survival of the fittest' and 'missing links'. There's something comfortable about all this traditional value, the familiar, putting things into neatly labelled boxes without thinking too much about what it all means. There was no attempt to bring these different spheres together.

A more holistic approach did not begin to emerge until the last quarter of the twentieth century. The interdisciplinary approach was called evolutionary biology, and it was to change our acceptance of Darwin's assumptions of linear evolution for good. While the laboratories and field locations continue to be occupied by particular specialists, one campus building has a broader clientele: the library. Where better to begin an interdisciplinary investigation into how evolution works than by collecting all the reliable knowledge together and analysing it mathematically and statistically?

The obvious contenders for the first prize at this task are David Raup and Jack Sepkoski, whose thesis of 26-million-year cycles of mass extinctions caused such a rumpus in the 1980s. After all, it was they who started all the talk of cycles, breaking away from the view of linear evolutionary progress. They were surprised and flattered by all the attention their 26-million-year-oscillations theory received, both the support from astronomers and the caution from paleontologists. Facing such strong opposition, there was no doubt about what they had to do to defend their position: get better data and better methods of analysis.

They had a plan to respond to the international uproar that the 26-million-year idea caused. Raup set about developing statistical tests to check on the validity of data about the ranges of fossils through geological time. They planned to expand the database of the first and last occurrences of all known species from the marine realm. Their 1984 speculation was derived from a large set of index cards giving details of fossils collected from marine rocks in North America. This had been built up over many years by many specialists, each with different perspectives and values. Clearly, if it was to be used as the basis of a grand theory, it had better be improved. But that's easier said than done. The more they thought about the quality and reliability of their data the harder it became to be convinced about the values those data contained and assumed.

One of their problems was to know at which level they should identify the specimens in their lists. Groups of species make up a genus, or several genera; further into the hierarchy, genera are within Families; then come orders, and the highest ranks of class and phyla. Genera are groups like *Homo* with a species *sapiens*, Families are on the scale of many genera, and phyla are big groups like Chordata. For our own species this hierarchy is often ranked like this:

phylum	Chordata
class	Mammalia
order	Primates
Family	Hominidae
genus	*Homo*
species	*sapiens*

Raup and Sepkoski knew that the pattern emerging from their analysis varied according to the rank in the hierarchy they selected. Which were they to choose? Not that they had much of a choice. Back in 1984 they had what they were given: a mixture of every rank, mostly with no thought about the real status of each of these divisions.

Jack Sepkoski moved into the library at the University of Chicago later that year. He decided to work at the level of genus and species, not just the Families and orders of their earlier list. It was a huge undertaking, requiring critical judgement as well as patience. The job was to take longer than he had expected, but if he was to bring he disciplines of geology, zoology and ecology together, it had to be done.

Differing rates of evolution

There was another good reason for Raup and Sepkoski to make a big new database. Their fellow paleontologists Niles Eldredge and Stephen Jay Gould had their own ideas about evolutionary mechanisms. They were based less on the analysis of large databases of first and last occurrences and more on the study of evolutionary trees built from hundreds of actual specimens of closely related species. In 1972 they first published their theory of Punctuated Equilibrium in *Nature*. Their idea is that evolutionary change takes place in relatively short bursts, separated by long periods of quiet.

These quiet phases are easy to observe in sections of rock strata where the same species is so often seen in sediments several metres deep. It means that the same species went unchanged for the length of time the rock took to sediment, maybe ten million years or more. Then suddenly a new species is seen in the newer rock above and continues once again for millions of years. A question is, where did that new species first appear? That's something about which we can't expect evidence because there isn't a tape recording of the action. The question is always evaded.

The first evidence for the theory came from Eldredge's study of trilobites, which looked like giant woodlice, segmented animals that flourished in the seas of the Paleozoic. They are good fossils to look at when thinking about evolution because there are lots of them, and each feature used to define the species is reliably stable. For example, tiny lenses in their eyes (which fossilise because they were made of the mineral calcite) show very characteristic patterns from one species to another, and Eldredge was able to show that these changed every couple of million years across North America. But through that time the species did not appear to change.

Furthermore, it is the rate of change that is different and important, because such waves of change, or punctuations, are found throughout the fossil record. For instance, the brain sizes of early humans change in waves of different sizes separated by stasis over long periods of time. The argument also emphasises the point that the rate is different in different species. Gould and Eldredge focused on the form and structure and specialised features of particular extinct groups. The features vary from one group to another, and the trends cannot be fitted into a common pattern. Their theory was devised to use different evidence from that coming from temporal databases like those my research group finds so interesting.

As with many changes through geological time, these punctuations appear to show up at different frequencies. It is reminiscent of Bak's sand-pile experiment. The 2-million-year cycles for North American trilobite eye patterns are surrounded by other punctuations occurring at different times. The structural features that change also have different levels of importance and can be expected to display the features of a power law – a few big changes and lots of little ones. Those changes that occurred during the Big Five mass-extinction events were massive. In 1993 Gould and Eldredge celebrated their theory having 'come of age' with a six-page review article in *Nature*, a sure sign that it is well accepted by the paleontological establishment.

That same year, Jack Sepkoski came out of the library with the first announcement of his new work. For a decade since 1984 the flurry of excitement about cycles of extinction every 26 million years had echoed around. In between the scores of meetings defending and elaborating that idea, Sepkoski collected and checked a lot of data. He found documentation for a quarter of a million fossil species of marine animals, less than 5 per cent of all that had existed since life began. Jack continued to defend the periodicity theory with his new database, larger and refined, but support from elsewhere was not forthcoming.

A decade for databases

Sepkoski's new database was to be as important for paleontology as floras are for botanists and identification keys for mineralogists. Unfortunately the data come only from the marine realm, and they are not generally available to the whole scientific community. What we need for fossils is something like what the Dictionary of National Biography started to do for historic celebrities. For each entry details are given of dates, places, relatives and descriptions. The entries for each should say what they're good at, how well they get on with others, and what special whims and fancies they may possess. As a result the data can be classified in many different ways, useful to a variety of different types of searches. You can work out family or professional relationships, see which are restricted to particular environments, and work out hidden trends and changes through time. If you've got good analytical software you see a lot of these trends and changes.

In 1991 I was contacted by the vertebrate paleontologist Mike Benton, a professor at the University of Bristol. Later he was to become the chief science adviser for the BBC programme *Walking with Dinosaurs*. We agreed that I would contribute to an international project giving the time of origin and extinction of all animal and plant Families. With about a hundred other specialists we were to check the entire scientific literature for the most reliable dates in geological times. It's now published as *The Fossil Record 2*. It was to become a public domain version of Sepkoski's private files of marine animals but it works only at the Family level and has no detail about the regions of occurrence.

Almost a quarter of a century earlier the editors of *The Fossil Record 1* had made the first attempt to present an overview of the history of life. They listed the large groups of fossil animals and plants that were then known. Unfortunately there was no strict methodology and some were listed at the Family level, others as orders, and many were known to be out of date quite soon after publication. So the database wasn't used to find patterns of diversification and extinction. There were to be many changes in the second edition, as much to do with new interpretations as with new knowledge, the creation of new genus and Family groupings. For example the 1972 first edition listed a total of 2,924 extinct Families of animals and plants. In 1993 there were 7,189. Most extremely, for insect Families the respective figures are 98 and 1,083. However, the fossil record is still patchy, incomplete and sometimes incorrect. The pictures that are painted vary, depending on whether it is the species level, or the genus, the Family or higher. Nevertheless, although you have to be careful how you use the data there are clear signals of evolutionary change.

All databases have their own weaknesses. One involves how the entrants are selected. A lot of the information is bound to be inaccurate, clouded with bad dates and names and with lots of gaps and unknowns. Though the users are constantly being warned of the shortcomings, many users wrongly accept the content as reliable and authoritative. A cautious sceptic was Sepkoski's colleague at Chicago, David Raup, who devised several ways of validating and checking the data. But problems remain and the dirty data are still impossible to clean properly.

The Fossil Record 2 is the most complete database we've got, with categories for marine, terrestrial, aerial and brackish ecosystems, and it's freely available. You can actually use the fossil database at *www.biodiversity.org.uk/fossilrecord2* and make your own selection to plot origins, extinctions and total diversification curves for all known Families of animals, plants, mushrooms and bugs. The dinosaur extinction curve in figure 2.3 is compiled in this way.

To many evolutionary biologists, such a large database is one of the most useful tools. Just as finding a beautifully preserved complete specimen excites a paleontologist looking for fossils in a quarry, so lots of good data stimulate people like Sepkoski to new heights of endeavour. However, finding a good database is about as hard as finding a good fossil. Then there's the problem of checking its accuracy and working out what methods to use for its analysis. The data have to be in a format that can be entered into some kind of presentational package and they must be publicly accountable as valid bits of scientific fact.

Through my own searches for the right kind of detail I've had plenty of disappointment; what sound like simple requirements turn into nightmares of administration and eventual bureaucracy.

In 1993 I was at the Komarov Botanical Institute in the newly renamed city of St Petersburg, in Russia. The roof leaked and there was no electricity or heating. I was organising a NATO workshop about climate change in the Arctic, and the Russian biologists were coming to London, joining others from North America and Europe. We sat in our overcoats and scarves being warmed by vodka, talking about plant evolution. On one occasion, in walked a grand old man. All through his working life his main project had been to record the reliable finds of plant fossils within the former Soviet Union. He recorded them on thin slips of cheap paper, handwritten in pencil. There are hundreds of thousands of these, with names, locations and geological age. To transcribe these data to a usable format would cost a fortune, which the new Russia is unlikely to find.

The same problem with money came later that year when I visited the Smithsonian Institution, just between Congress and the White House in Washington DC. As you can imagine, it was very different from the Komarov and there was no vodka. I came across their 850-drawer index-card cabinet of fossil plant occurrences from all over North America. This time they are typewritten on cards, just as expensive to transcribe onto disk. So again I left without any of the valuable data.

On my arrival back in the UK, I received a letter from Massachusetts. For more than thirty years an American oil exploration company had computerised data of fossil occurrences described in the scientific literature. They did it first on cards, then tape and now disk. The database has more than a million records of occurrences, locations and ages. All these data were drawn from the public work of academic scientists, rather than researchers working in industry. Knowing the information came from the public domain my research group gaily put it on the internet. We were rather proud of the way you could very easily search for names, ages and places with fast responses. Within a few weeks I received a very official letter from a Boston lawyer. The letter threatened me with all manner of nasty punishments if I didn't remove the database from the web immediately. It turned out that according to the international law of intellectual property they were in the right. The collection of data was theirs, even though the original bits came from the published domain of academia.

These three incidents show the immensity of the obstacles to creating and accessing large sets of data electronically. Technical, political, financial and legal obstacles are hard to overcome, so large sets of data about evolutionary and environmental changes are precious. Understandably, some academics are reluctant to share their own hard-won resources, afraid that others will steal their thunder and make an easy reputation. These are trivial incidents but not unusual among the daily difficulties to meet the urgent need to monitor change to environment and life on our planet.

A new shape to evolution: exponential change

The Fossil Record 2 was eventually published in 1993. The database offered a challenge to those interested in finding patterns in large data sets and also presented a standard against which specialists can compare their own specimens and ideas. It could not have appeared at a better time, presenting new data to test the two controversial theories from American paleontologists: Raup and Sepkoski's 26-million-year oscillations and Gould and Eldredge's Punctuated Equilibria. With the hint of success at sorting out the first diversification of angiosperms at the Cenomanian–Turonian boundary my research group was ready to take on a much bigger case. Could we help solve a central mystery of evolutionary theory? Does evolution happen gradually, in steps, or by some other routine?

The search for an answer to this question is rather like groups of private detectives competing to find clues to solve a case. They are international groups, each with their favoured techniques and attitude. The Americans like to check the basic building blocks, the data themselves. The French are very concerned that a logical protocol has been followed and that the analysis of data is properly validated statistically. The English have hunches, a suspicion of simplicity and the desire to test. All three groups of scientists are keen to prove the other wrong and maybe won't be too upset if they are eventually proved to be wrong themselves.

Taking editor's privilege, Mike Benton was the first to use the data we had all brought together, looking for signs of the shape of evolution that Darwin had assumed. His article published in *Science* shows lots of curves with different changes in Family diversity through time, the rates of extinction and the number of Family origins. They show trends for continental and marine organisms, many high peaks and no clear patterns. But one thing is clear: the rate of Family evolution was not gradual. None of the basic curves were anything like a straight line, so Darwin's assumption of evolution happening smoothly was immediately ruled out. The peaks at period boundaries are too steep for that, and there are plenty of other changes going on at different times and different places. The curves look like a lot of chaotic mess. Neither were there any oscillations on a regular timescale, let alone the hypothetical 26-million-year interval so much favoured by Raup and Sepkoski.

Nevertheless the plots of numbers of Families do show the Big Five mass extinctions as clearly as Sepkoski's own curve (see figure 2.4). At least that means that the two databases confirm the same trends – good news for the reliability of the information. It may also mean that the mass-extinction events are equivalent to Gould and Eldredge's punctuation. Their original idea was based on data entirely from marine invertebrates, while Benton's plots derive from data based on continental organisms as well. Other plots of total extinctions confirm the Big Five as well as other more minor events, and show delayed origins of new Families.

Many of Benton's curves contained steps of different sizes, some went down as well as up, but the clear trend was of increasing diversification through geological time. Could the curves give some support for step-like patterns of diversification? The evidence was ambiguous, open to creative interpretation by supporters and critics alike. It was clear that the new database was a powerful resource upon which to test ideas about the history of life.

But there is a discrepancy between the two very different kinds of evidence available. Some is structural and the remainder analytical. Are we wrong to use them both to solve the same problem of how evolution works? Gould and Eldredge's theory is based primarily on things like their observations of trilobite eye characters. Benton's curves are based on the number of Families of all organisms that are present through time. Why should the two variables be expected to show the same trends? It looked as though the case for finding what runs evolution was still wide open.

In 1996 more favourable evidence to support the theory of punctuated equilibrium came from an unexpected source. It was from France, and the geophysicists Vincent Courtillot and Yves Gaudemar, eager to put the new data through their very sophisticated mathematical procedures. They had a new technique designed to detect patterns within very large sets of data and they had used it to find hidden meanings in *The Fossil Record 2*.

Their calculations detected a sudden rise in the middle of complex and apparently featureless data. What's more, once a peak is reached the same high level of diversity is maintained at that plateau. The results strongly supported Punctuated Equilibrium and hint that most of the peaks come at times of mass extinctions. Each step comprises a gradual start, then a rapid rise in the rate of diversification, and finally there is a static equilibrium, typical of punctuated step-wise diversification. In the Family data there were also lots of smaller steps in between, like the minor avalanches on Bak's sand pile. Each step comprised a gradual start, then a rapid rise in the rate of diversification, and finally a static equilibrium. This was interpreted as typical punctuated step-wise diversification.

With support from these heavyweight scientists it suddenly looked as though the last century was going to end with punctuated evolution the sound favourite to explain biological diversification. And it seemed that *The Fossil Record 2* was providing the final proof.

81

But Mike Benton had had a hunch that he briefly mentioned in his summary in *Science* magazine. It was based on the observation that the overall increase in Family diversity from the database shows a power law. It is the same process that we saw in figure 3.2 from Bak's counts of the number of sand grains in avalanches within the sand pile. There are very few large avalanches or changes in Family diversification, and many more smaller ones. That's because most evolutionary changes are small adaptations in morphology and larger-scale changes happen much more rarely. If this hunch turned out to be correct, it would have significant implications for our understanding of evolution. It would mean that evolution follows an exponential diversification, rather than a linear or punctuated one.

A well-known example of exponential change is found in accounts of rises in species populations, where numbers within a species increase at ever-faster rates until the graph's line reaches almost the upright vertical. The mathematics of logarithms won't allow that point to be reached, but usually some disturbance to the system stops it going on up. So, in the case of the Big Five mass extinctions (figure 2.4) and the exponential change in total diversity (see figure 3.5), extinctions caused by changes outside the system prevent the curve rising far to the vertical. It is a characteristic of life evolving as a self-organised system, with avalanches like a sand pile, some from within and others from outside forces. For example, changes inside may be such as genetical recombinations and small structural improvements. From outside, climate change and new competitors can upset the balance. All these can cause the loss of a species. Life on Earth needs extinctions for it to change and diversify.

Most large groups seem to double their overall diversity in the first part of their existence, once the group is established ecologically. This would happen if the change was exponential, but there was no clear evidence from Benton's curves. What are needed are statistical tests on the data to be sure of the shape of the changes. For example, the analysis of separate sets of ecologically similar phyla and the clear separation of marine and terrestrial components would help. Further-

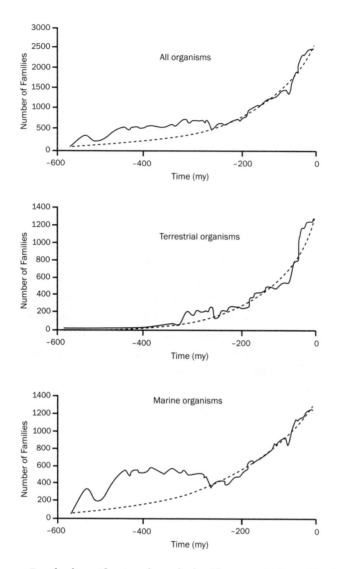

Fig. 3.5 Family diversification through the Phanerozoic from *The Fossil Record 2* database, for all organisms, all marine organisms and all terrestrial ones. Compare the marine curve to figure 2.4 after Sepkoski, 1993. The dotted curve is the calculated exponential rise in diversity. It rises almost to the vertical for the terrestrial Families.

more, the Americans like Sepkoski and the French such as Courtillot, let alone most of the other scientists involved, wanted deep quantitative evidence. There was one obvious place to look for support, obvious that is to those few evolutionary biologists who know that the sand-pile avalanches share mathematical properties with exponential curves. If sand on piles and traffic in jams can show features of exponential change, so may evolution. What's more, just as the motorway has a limit to its capacity for cars, so has the Earth for organisms. My small research group was ready to swing into action.

The first thing we did was to make the *Fossil Record 2* database inter-active on the internet (*www.biodiversity.org.uk/search/fossilrecord2.html*). We then wrote a computer program to separate the records from differ-ent ecological sources that were distinguished in the database: marine, lagoonal, freshwater, terrestrial and the airborne. These features had been allocated to all the Families mentioned in *The Fossil Record 2*. We also added a taxonomic hierarchy to each of the animal or plant Families: things like order, class, phylum and kingdom. This was introducing more controversy into the data, for nearly every specialist has his or her own scheme for these rankings. We had a similar problem with the age of each occurrence, whether they were the names of the stratigraphic divisions (see figure 1.2) or the numerical age. They vary from time to time according to different national schemes. The more complete the data, the more controversial is their content and the more debatable the conclusions from their analysis.

One advantage of these interactive databases on the internet is that you are not stuck with the range of output that the author wants you to see. You can choose your own variables and plot your own graphs. In the near future the impact of this facility on the role of the reader will be profound. No longer will we passively scan our eyes along the sentences. Instead, we will have to think about which variables to select and then construct a range of outputs that best relate to our own particular interests. It won't be easy; much harder work than just reading and being fed by the author.

Working like this, we selected all the marine invertebrate Families in the database. The curve (figure 3.5) was like Benton's as well as Sepkoski's. You can recognise the same peaks and troughs, the same Big Five extinctions. They are the same shape, with the same big rise in the Paleozoic, ending at the P–Tr extinction event. Because the database has Family diversity from terrestrial environments as well, we get that equivalent curve, except that it lacks the Paleozoic rise, which happened as life made its move on to land. The same shape started 320 million years ago and the last three of the Big Five are clear. Although self-organisation maintains an overall exponential pathway, the Big Five externally-induced events do have an effect. They have the effect of pushing the curve to the right, prolonging the rise to the vertical. So far, so good: the patterns were familiar and confirm the view of the Americans and the French.

We tried every conceivable way of selecting different data sets and testing the outputs against the theories that we knew. One particularly interesting kind of selection was to lump together all marine organisms and plot them separately from terrestrial ones. Quite different eco-logical conditions control each environment, and this shows up in the different-shaped curves of figure 3.5. There is a huge literature relevant to so many of the different outputs, and each time you select a different option you are entering a new set of academic controversies.

Our next test of our exponential model for diversification was to use a different database. The obvious choice was Sepkoski's marine invertebrate database, the result of his ten-year sojourn in the library. I wrote to ask if we could include it in our testing plans. Permission was not granted and no reason given. Instead I turned to my friends in the oil industry, who I knew were sitting on stacks and stacks of identifications of micro-organisms. They find these in the deep cores drilled when exploring for oil and they help tell the age of the rocks and whether the environment at the time of deposition was a likely source of oil. One of these databases had 3 million records.

If the quality of some of these records was suspect, there was no doubt about the quantity. We knew that with very large amounts of data the result of adding some of the variables together was to give very clear trends. It turned out to be so, with a smooth exponential curve starting slowly in the Jurassic and reaching close to the vertical in the latest Tertiary. It showed that Benton's hunch was supported from yet another source.

Now we started to be creative and went on to apply the ideas from self-organised systems. We began by looking for power laws and pink noise. We plotted the range of size of the extinction events in the database: there are fewer large extinction events and many smaller ones. That's no surprise: the P–Tr event is at the top of the league, the K–T is the next of the Big Five. Then there are more and more ever-smaller ones. That range of happenings typifies natural systems which give curves symbolising a power law and pink noise. It is absent from all unnatural time series.

Those days spent formatting data and testing new computer programs were especially tense in our research room. A clear power-law curve was another sign of excitement coming from our new analysis: a simple conversion of the total number of Family occurrences into a logarithmic value, plotted every one million years. No straight line suddenly flashing up onto a computer screen has attracted more excitement – there was a roar of approval. So we had an exponential and a power law to support our idea that the rate of evolution was increasing steadily towards infinity.

The next task was to test the data for signs of pink noise, the final confirmation of the Benton hunch. Our specially formatted data were in the same Microsoft Excel file, but the program doing the number-crunching was new. Dilshat had struggled with the program long and hard, firstly writing it and then smoothing out the bugs. Eventually he poured the data from the Excel file into his software and out popped the curve (figure 3.3) that shows that the system features pink noise.

It's times like this that make the hard slog of being a professional scientist all worth while. Despite the hard business of learning, the exams, part-time contracts and bad pay, there are still dozens of bright applicants for every research job. They do it for the chance of this excitement, which is hard to find elsewhere in the modern world. The curves were spread all over the tables and benches in our office. More copies of different versions confirmed the main trends. We stayed till late, called in at the shop next door for a bottle of champagne, re-checked the results the next day; e-mailed academic friends to check the main ideas; read a tenth time the cautionary arguments from our opponents, Courtillot, Gould, and Kirchner and Weil, to make sure that we had covered ourselves. We were ready to go public.

The manuscript that Dilshat e-mailed me about in Taiwan gave results from this analysis of *The Fossil Record 2*. The results suggest that the fossil record is a self-organised system and that evolutionary change happens exponentially. We can add biological evolution to Per Bak's list of things that are in control of themselves and change by forces from within. For us, the power and control of change comes from within a complete and large system. All the complexity is within the one huge machine, whether sand pile or Earth. A famous icon of the last century was the NASA picture of the Earth taken from the moon. Seeing the whole machine from outside the system is one of the most beautiful of images.

4

From Dinosaurs to Us

Recovery from Chicxulub

It's likely that the dark dawn of the new Tertiary period 65 million years ago lasted at most a year or two. At first the clouds of extraterrestrial dust as well as organic and inorganic debris from the Earth formed their own interactions and patterns of circulation. There were aerial explosions, and complex reactions to the impact came from within the Earth itself. The fires and violent storms soon lost their energy and the clouds began to get thinner. The atmosphere was still dark and cold, a silence interrupted by gales of frozen rain.

In the continuing gloom, all to be seen was through the light of the dying fires. I doubt that the Earth has ever been bleaker than during those years, when most living things survived in trauma and in hiding. The first positive sign of recovery came as the sunlight began to break through the dense atmosphere and once more shine upon the Earth. The low productivity in the oceans that had been maintained for many months suddenly went on to increase quickly. The hell was over.

The first days of sunshine after the K–T clouds were celebrated by spontaneous biological reactions. The rapid return to photosynthetic activity caused a great wave of evolution within the many tiny phytoplankton. Sea level was low and a new life returned in the form of new species of plankton, molluscs and fish, though we are very short of good details. A similar abundance of microfossils laid down in the oceans both before and after the boundary clay suggests that the balance of life there returned to normal quickly, though with very different biodiversity. Although the ammonites were completely extinct and although the species of other major groups were replaced, the world's oceans show a smooth transition through the crisis from one fauna and flora to another. Nearly all these new species of plankton, molluscs and fish were of genera established during the Cretaceous.

Evidence now shows that the environmental changes caused by the collision of the Yucatán meteorite were restricted to a very short length of geological time. We think that the whole catastrophe lasted only 10,000 years, with many details of the event preserved in the deposit of iridium layer which I described in chapter 2. When interplanetary dust and iridium and the toxic debris from the fires themselves settled over the Earth's surface between the Cretaceous and the Tertiary, a layer of these particles sedimented where it could. It was the same layer that Walter and Luis Alvarez first made famous just a quarter of a century ago with their discovery.

Elsewhere, the Earth became a much quieter place than at some of the other mass-extinction events, when the sea experienced long crises from other sources such as volcanic activity. At the Permian–Triassic (P–Tr) boundary, 245 million years ago, prolonged lack of oxygen, high temperatures and acid rain caused much more havoc. As an event within a self-organised system, this Cretaceous–Tertiary (K–T) ava-lanche, 65 million years ago, was small enough to allow the sand pile to return to its earlier shape surprisingly quickly. The sand pile of evolutionary biology continued to build from its own internal forces, despite the big kick from the Chicxulub asteroid. That is shown by *The Fossil Record 2* and other data to be interference from outside the system, a kick to the sand pile. Other scientists such as Gould and Eldredge, who support the step-wise Punctuated Equilibria for the evolutionary process, see it as another advance up the slippery pole of evolution.

Many organisms were well protected from the mayhem and show no signs of damage or change. They continued their former lives unaffected by the environmental change. On the other hand, when the ecosystem was more disturbed, with frequent upsets as a consequence of the catastrophe, then recuperation was slow, with altered ecosystems and new ecological relationships. These alterations became global and are the most prominent legacy of the dinosaur and ammonite extinctions.

But other questions remain. How did the entire biosphere, all the biology and ecology, respond to the extinctions of large organisms like ammonites and dinosaurs? The changes in fauna and flora had effects on the way living organisms were recovering from the traumas of finding themselves with new neighbours, let alone new surroundings. The effect was dramatic for some forms of life while it made little or no impression on others.

Nevertheless, numerous new species and genera evolved just after the K–T boundary at the beginning of the Tertiary period. Paleontology textbooks have lots of drawings speculating from skeletal evidence on what the new furry mammals looked like. Their appearance may be uncertain, but there is evidence to show that there were lots of them, new species as well as growing populations of individuals. New Families of land invertebrates, freshwater fish and even lizards took over the free space left by the greedy dinosaurs. Marine fish and plankton diversified just as quickly at the species level, but there was no immediate change in the Family constituency or the individuals' behaviour.

Recovery from the catastrophe in the marine realm was controlled by how and when oxygen returned to the water. There is some debate about what happened. Some of the small shelly zooplankton seem to have lived right through the catastrophe, not to be replaced until there was full recovery. Other kinds of plankton from deeper ocean show a range of different responses. Some groups became extinct, others survived, and the majority radiated as new species, genera and even Families. What's more, these responses were immediate, and we can see them happening within a few hundred years of one another just after the event itself.

The trend is clear from other kinds of evidence that the K–T event had a short duration. Isotopes of helium, for example, show that the rates at which sediments were deposited returned to normal after just a few thousand years. More evidence is becoming available from cores being drilled from the deep sea, giving more clues for these plankton specialists, geophysicists and sediment chemists, so the controversies will rumble on.

More traditional signs of recovery from fossils of larger sea creatures such as bivalves and fish are fewer. There are not so many locations of the right age explored and most lack good specimens. What have been found turn out to be quite new species of Cretaceous genera, which were then rapidly replaced by new genera fossilised in rocks just a few million years younger. Though it's not yet precise, the trend is clear: the K–T event was followed by a major change in marine biodiversity. There is also evidence that the same radical diversity occurred on land and in the air.

Of all the curves of Family diversification that my research group has made from our analysis of the *Fossil Record 2* data, I think the most amazing so far is from the birds just after the K–T boundary. Plotted on their own the bird Families show a very low diversity throughout the Cretaceous. Like the mammals the group never really succeeded, presumably because the dinosaurs kept them in check. Sixty-five million years ago both mammals and birds exploded in their diversification. A most exciting thing is shown when the Families of birds and the Families of dinosaurs are plotted together as a single curve (see figure 4.1). The K–T extinction is clear. So is the following diversification of birds. But together, from the early Mesozoic and with the K–T recovery, the one curve assumes the shape of an exponential.

Some vertebrate paleontologists have been arguing fiercely for years about the evolutionary relationship between dinosaurs and birds. The arguments began with the discovery of feathered *Archaeopteryx* from the Jurassic, and now more new feathered dinosaurs are being discovered in China and compared with American specimens. The debate is likely to be resolved by studying feathers and bone mechanics and structure. It remains to be seen whether the exponential shape to the curve in figure 4.1 has any significance for these comparisons, or for more general arguments about evolution. Nevertheless it is striking to note the shape

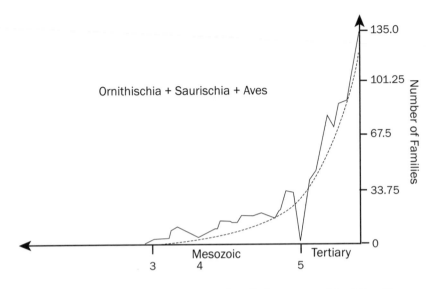

Fig. 4.1 Family diversification curve for all dinosaur and bird Families whose time ranges are listed in *The Fossil Record 2*. The dotted line is the exponential model for the curve. The last three of the Big Five mass-extinction events are numbered on the horizontal axis. Extinction 5 at the K–T boundary saw the reduction of dinosaur Families to zero and the rise of the bird Families back to the levels of the exponential model. (original compilation)

of the joint curve, support for an exponential pathway for evolution and for a close relationship between dinosaurs and birds.

Mammal Families were also increasing rapidly early in the Paleocene, achieving record numbers of new species, new Families and individuals. The new groups included Carnivora, Insectivora, Primates and rodents and quickly spread into the new forests of the northern hemisphere and then moved southwards. They were also increasing in size, and there was no shortage of food and clement habitats. They were living through times of great wealth and tranquillity for more than the first 20 million years of the Tertiary period.

The other large terrestrial group to radiate quickly once the planet had recovered from environmental devastation was the flowering plants. They were evolving as never before, and they were changing in tune with the mammals. For both groups, these early species and most of their genera are now extinct, but the plant Families that were becoming established then are nearly all with us still.

Most plants seem to have survived the K–T impact to make a good recovery, presumably through the well-defended structure and position of their roots. Photosynthesis stopped, trunks burnt down and leaves didn't shoot, but roots that stayed quiet for decades did rejuvenate. It's not until you look hard at the evidence that you see just a few subtle changes. The strongest changes in vegetation show up across the boundary, close to the site of the K–T impact in western North America. A few of the Cretaceous plants such as the normapolles and aquilapolles I mentioned in chapter 2 stop growing in their home region; either they became restricted to outlying pockets prior to extinction, or they migrated to find old ecosystems in the far north. There, they eventually became extinct.

One of the easiest ways to examine changes in vegetation across the boundary is to count the different quantities of pollen and spores accumulating in every one or two millimetres of sediment, each representing more than a thousand years of accumulation. For thirty years from the 1960s thousands of botany or geology graduates like myself trained in these skills of microscopic examination. We can identify what grew near to the sampling site, reconstruct ancient ecosystems, and help identify the age of the sediments. From this kind of information these specialists reconstruct the original environmental conditions and help restore more ancient landscapes and ecosystems.

Hundreds of thousands of pollen and spores are usually preserved in a few millimetres' thickness of sediment, each giving a clear identity of their parent plant. Diagrams of the results from counts just above the ash layer in North America and Europe show lots of fern spores and little else. This makes sense, because after forest fires, ferns are usually the first colonisers. You also find them just after volcanic eruptions, and they were first to grow in European bombsites after the Second World War. They like the ash in the soil and they have no problems with low-light regimes.

Huge databases have been created from this work on pollen and spores, and my research group is using some of them to test our ideas of evolutionary biology. From these data we looked at the changes in the fossil pollen through the last 65 million years, and to check their validity we set the patterns emerging against those from very different disciplines such as sedimentology and climatology.

The pollen databases confirm the rise of deciduous trees and shrubs, especially broad-leaved trees such as oak and lime. These plants also diversified as species and genera 65 to 55 million years ago, during the Paleocene (see figure 1.2). The same interval saw the beginning of the great mixed conifer and broadleaf forests that began to cover the higher latitudes of the northern hemisphere. In other parts of the world there is much slower diversification in early tropical rainforest, warm-temperate floras, and in the more temperate ecosystems in the south. All the evidence points to a global expansion of forests at the time, in the number both of species and of individual shrubs and trees in these vast terrestrial ecosystems. It was the time of the highest growth in biodiversity that the planet has ever experienced.

Age of the broadleaf forest

Global environments were set on a course of very steady change for 25 million years after the extinction of the dinosaurs, having recovered from the catastrophe. The overall trend was dominated by an increase in global surface temperatures. The band of the tropics around the Equator slowly widened out towards the north and south, reaching a maximum extreme about 50°N, 45 million years ago. Now it is 15°. The cause of this slow and steady increase in gently oscillating temperature is not too clear, and the position of the planet in relation to the sun and other systems may have had some influence. With changing sea level, the larger landmasses encouraged warmer summers and without ice at the poles, continental winters would have been much warmer than now. One unusual heat source came from the release of huge quantities of natural gas buried in continental shelves that caused sporadic explosions on the surface. These blowouts burned for thousands of years and caused regular erratic disturbances in the environment, possibly causing a peak of global warming at the end of the Paleocene. Within the overall warming trend, there were more and more local ecological changes influenced by alterations in the global environment. Ecology and the geographic distribution of animals and plants were becoming more varied than ever before.

The newly evolved mammal groups diversified fast in these new ecosystems, able to take advantage of the new environments without threat from the old predators. They lived in the newly established broadleaf forests that provided good protection from their new enemies, mainly other mammal species, and whose trees also provided leafy food from sunlit branches.

The new flowering plants took a liking to this space once it was vacated by the giant reptiles and shady ferns. It was filled very quickly, and trees with large deciduous leaves flourished in the wet and warm climates. Some of the Cretaceous conifers survived by adapting to their new neighbourhoods, since they were tolerant of the warm, wet conditions. Others retreated to drier parts of the world, where they remain in small numbers as rare species of pine. Most *Pinus* species are still subtropical and survive only in very small numbers, quite the opposite of the popular perception of cold snowswept forests in the far north monotonously populated by pine, birch, ferns, moss, and very little else.

Just as the Jurassic was the Age of the Dinosaurs so the Early Tertiary was the Age of the Broadleaf Forests, with big trees dominant, some evergreen conifers mixed in, and a wide diversity of species and genera. There were oaks and planes, sycamores and hickories, as diverse as the specimens planted in an arboretum. Among the conifers were the sequoias, umbrella trees, Douglas fir and spruces; many of the species became extinct but the vegetation would look familiar to us today. These grand trees were living through times of warm climates and rich environments. To the south tropical rainforests expanded to cover regions as far north as Paris and London while huge redwood forests had migrated to Vermont, Berlin and Beijing. They were ideal territory to encourage birds and insects to diversify, while other invertebrates occupied the new specialised niches opening up in these new environments.

As well as these warm climates through the Early Tertiary there was more temperate vegetation to the far north. The flora is like a dream for botanists: extinct species, big leaves and tree trunks, but very familiar genera and very familiar plants: giant birch and plane trees, productive vines and big thick shrubs of alder and holly, all growing densely together. In the clearings were ferns and heather looking very familiar but big, healthy, lush and aggressive. Many of these species don't grow any more, partly because the ecosystem hasn't existed since the poles became glaciated. The temperate deciduous trees had the largest leaves because the summer sun was low in the sky, giving less energy for photosynthesis.

To the south the forests began to look more and more like modern tropical and warm-temperate ecosystems, and though the species have become extinct, most of the major evolutionary branches of plants and animals were formed in these Early Tertiary forests 55 to 45 million years ago. In the northern hemisphere a single landmass had comprised the present North America to the west, Europe in the centre and Asia to the east. Tropical mangrove swamp occupied most of these shores, and the newly forming tropical rainforest stretched for thousands of kilometres east and west. Rich, humid environments hosted wetland swamp cypress forest like that in modern Florida, and upland forests with multistoreyed ecosystems, lianas, the smell of rotting vegetation and the hum of insects and birds.

Often the forest trees were close together, making dark refuges where only the high leaves caught the sunlight for photosynthesis. There was very likely a particular kind of biochemistry going on in these leaves because there was a lot of CO_2 in the atmosphere, much more than today. The greenhouse effect was at its strongest 55 million years ago, which is at least one reason why the climate became very much warmer at that time. We know that this was so from many different kinds of evidence. One of the strangest is from the study of the density of stomata, the pores on the lower surface of leaves through which gases pass at a controllable rate. The greater the number of pores the faster the transfer of gases, and therefore the greater the rate of photosynthesis, which means the greater the concentration of CO_2. The forests thrived, reaching their own high levels of biodiversity and population density.

But the forests' success restricted their ecological structure to shrubs and high-tiered foliage. Very few herbaceous plants were growing on the dark forest floor as a result. These shaded lower tiers were home to ferns and mosses. The more humid the warm closed atmosphere, the better it was for these plants. This kind of vegetation of big warm-temperate forest, not well known today outside the tropics, had been very abundant since the forests of the Jurassic and Cretaceous.

The mammals continued to diversify through the Age of the Broadleaf Forest, expanding the range by around 20 to 45 Families through those 10 million years. It was the fastest-ever increase in the origin of mammal Families, and several thousand new species came into existence. They were mostly small, few bigger than an average-sized dog, but each new species had a larger body mass than the earlier species of the same genus. A trend of increasing body size for Tertiary mammals was quickly established, as there was no longer need to keep out of the way of dinosaurs by being tiny.

Through the time span of the Early Paleocene, 65–60 million years ago, the same mammals got bigger and bigger, each new species adapting new structures and physiology to cope with the increased weight. Gradually some of them developed a browsing habit, sticking their necks upwards to seek the lush leaves from the juicy new angiosperms. But for some reason it was to take many more millions of years before the mammals were able to grow big enough to reach the tree tops, the early giraffes not originating until the Oligocene, 35 million years ago. The reason for this delay is still a mystery, for the prize of choice fresh leaves around the crowns of high trees must have been very desirable.

The Atlantic River

Before the end of the Paleocene, 58 million years ago, a few kilometres to the south-west of where the Tirefour Broch is today, there was sudden mayhem. For 7 million years the small mammals and the thick forests had been slowly changing through times of a deep peace. Though it continued to get warmer, the environment had shown no signs of big changes. The Atlantic Ocean reached no further north than southern Europe and Florida, the continents being connected by Greenland. Suddenly, a great line of volcanoes began erupting along what is now the mid-Atlantic ridge (see figure 2.5) from the Azores, through Northern Ireland, the Isles of Mull and Skye, the Faeroes, up to northern Greenland. It was time for another avalanche on the sand pile, this time from forces within the Earth system. The mayhem was targeted at just one region on the globe and has quite different effects from what we have learnt to expect from mass-extinction events.

Environmental change was being stimulated by drifting continents, forced by the sea floor spreading new Earth's crust out from deep within the mid-Atlantic ridge (see figure 2.5). The continents began to move apart, forming the northern Atlantic Ocean. The movements were to push what is now Europe away from Greenland and North America at a rate of a few centimetres a year. As the Atlantic River slowly widened and as it unzipped northwards the new seaway slowly began to connect the southern Atlantic Ocean to the northern Arctic Ocean. Although these changes to the environment are much slower than mass-extinction events they can have massive impacts on the evolutionary changes of species and communities.

I first visited the Isle of Mull on the west coast of Scotland in the 1960s when I was a research student, searching for fossil plants halfway up a cliff on the south-west coast of the island. It is one of the most beautiful parts of the British Isles, isolated and perfectly quiet. The weather is either tranquil or violent, and often very wet. On the shore during that first visit there was a circular metal buoy 5 metres in diameter, very battered and rusty, presumably beached in a storm. In big letters it announced 'Wood's Hole Oceanographic Institute', a famous centre for marine environmental and climatic research south of Boston in Massachusetts. It's where marine phytoplankton are studied, though plenty else besides – weather patterns, climate change, global currents. The Gulf Stream had brought their buoy 5,000km away across the sea to this beach.

Next day, we were looking for fossils on the south side of Mull. To get there, the map told us to pass through a village called Shiaba, situated strangely far away from any road, at the end of a dirt track that wound beside peat wetlands. The path led past crumbling dry-stone walls to a tumbledown collection of shallow ruins leading off from a central pathway which had surely been the main street, now grassed over and let to grazing. Erratic building stones were scattered every-where, just a few shallow walls of the rooms being discernible. One night at the height of the Clearances in the late eighteenth century, it has been recorded, the whole population of this tiny village walked down to the shore and boarded a boat for North America. For these victims of Shiaba's poverty the move to the New World offered a release from death.

Fifty-eight million years ago there wasn't a beach, let alone an Atlantic Ocean, just a line of volcanoes over 5,000km long stretching from south to north, belching ash and smoke and fire between Scotland, Ireland and Norway in the east and Greenland in the west. The ocean slowly opened, a few centimetres a year, northwards like a zipper. What had been a simple land connection for animals and plants from Europe across to America through Greenland was coming to an end, and the land bridge between them was finally broken about 30 million years ago. These slow geographical changes influenced both the climate and the biology. They affected migration patterns, weather patterns, the way the whole system moved and interacted. Change one feature

in a self-organised system and you get a reaction in many other dependent parts of the whole complex machine.

The fossil plants that have been found on that Mull coastal section
show what the vegetation was like when the zipper started to open 58
million years ago near the end of the Paleocene. The evidence of
continental drift is trapped in the basaltic outpourings from the volcanoes, which stretched at that time from Antrim, past Mull to the
Faeroes. As it poured down into the sea the quickly cooling molten
rock contracted into long columns sometimes as tall as four-storeyed
houses. Each column was about half a metre in diameter, with a
hexagonal or pentagonal outline. Between the volcanic lava flows which
formed these columns there are interbasaltic sediments from the quiet
intervals between eruptions.

Each of the 120 flows calculated represents thousands of years of
high volcanic activity. Between the violent periods, the environment
went through the same sequences of ecological change and recovered.
Once the lava flows had solidified, the smoke and ash had settled and
the storms had washed this bleak space clean, the living systems returned
to reach the same stable state. The forest grew on the thin volcanic
soil at each interbasaltic interval. There were big trees like planes,
beech, shrubs of alder and lianas of vines. Other trees were *Ginkgo*
and evergreen pines and redwood-like giants of *Sequoia*. In the warm
humid undergrowth ferns flourished as well as mosses.

We know this flora from a wonderful collection of fossil leaves
collected in the 1880s by Starkie Gardner, a Scottish amateur paleontologist. For a living he made gates; those at Holyrood Palace and at
Victoria Gate in Hyde Park were his work. Yet in 1884 Gardner
received a grant from the British Association for the Advancement of
Science to dynamite the cliff section on Mull. The condition was that
half of the fossils go to the Natural History Museum and the other to
the landowner. This was the chief of the Campbell clan, the Duke of
Argyll, who insisted on demanding money from whichever institution
would pay for his half of the collection.

Museums at that time expected voluntary donations of national heritage. There was no way they would pay Argyll's price. The Glasgow and Edinburgh museums had very little money for that kind of thing, while the richer British Museum had already taken its half of the collection. In anger, without a buyer and with the threat of more of the fossils going to England, the chief of the Campbells ordered that the priceless rocks be used as hard core for roads on his estate. Nevertheless we know enough from what does remain, mostly at the Natural History Museum in London, to be able to compare these specimens with fossils from Antrim, the Faeroes, Greenland and Spitsbergen. We call it the Brito-Arctic Igneous Province.

Fifty-five million years ago, when the volcanoes in the Brito-Arctic Igneous Province stopped erupting, ferns were the first to colonise the newly cooled volcanic soil, their spores blown over the lava fields from nearby forest clearings. Shrubs were next, followed by birch and a few conifers. The climax forest of the stable state included the familiar broadleaf trees of plane, alder, birch and oak. With so little evidence we are still unsure where the animals fitted in, but many extinct species of small voles and hoofed mammals are known, mainly from their fossilised teeth.

It was much the same kind of pattern 54 million years later during the ice ages, glaciers spreading and retreating, the environment wrecked and then recovering. Sometimes the changes are cyclical, repeating the same sequence of changes in succession. Other changes are irretrievable, with no return to where they began. Whatever the cause – volcanoes, glaciers, humans – the ecological system recovers.

On Mull, at the end of the Paleocene 55 million years ago, these sequences of recovery and change in the vegetation happened tirelessly, and you can see the inter-basaltic plant sediments on the cliffs to the south-west near the hamlet of Bunessan, just before you get to the ferry that takes you to the island of Iona. Further across the water, exactly the same kind of basalt columns are between high and low tide at the Giant's Causeway in County Antrim (see figure 4.2). Between the two was the Isle of Staffa, with Fingal's Cave, carved by the waves within the same flows of molten rock deposited as pillars of columnar

Fig. 4.2 The Giant's Causeway on the coast of County Antrim is formed of columnar basalt. This came from eruptions along the mid-Atlantic ridge 55 million years ago when the North Atlantic Ocean zipped open northwards to the Arctic.

basalt by the same lava flows. And if you go to the right places, between the flows of lava you can find clay sediments from the quiet millennia between the eruptions. If you are really lucky you will find the sediments from the bottom of a lake, with the compressed fossils 58 millions years old.

The same extinct flora has been found at many of the locations where the inter-basaltic sediments survive: Mull, Antrim, offshore Rockall, the Faeroes, east and north Greenland, and Spitsbergen. These places were all part of the same landmass that the volcanic explosions were beginning to separate into two continents. At one stage in the subterranean action it was touch-and-go whether the division between the continents should go down the west side of what are now the British Isles, or the east. If it had gone the other side, Britain and Ireland would be part of North America, not Europe. Geographical groupings of future nations were settled 58 million years ago without consulting politicians.

To help understand more about this opening up of the northernmost North Atlantic Ocean, on 19 June 1985 *JOIDES Resolution*, the deep-sea drilling ship of the Ocean Drilling Program, left Bremerhaven to go to work in the Norwegian Sea. The ship made three holes, one on either side of the divide, and the third down to the basalt flows which came from the ridge itself. This last hole was in sea 1.3 kilometres deep and the cores were over 1,200 metres long. The drilling produced cores of sediment recording the complete history of the opening process. There were 120 cycles of volcanic eruption, some dated radiometrically to give their geological age. The whole period of volcanic activity is thought to have lasted less than 3 million years, so it follows that a cycle of eruption and the quiet period afterwards lasted on average for about 25,000 years.

The ship docked at St Johns, Newfoundland on 23 August, mission accomplished. But for my Norwegian colleague Svein Manum and myself, the work was just to begin. We were commissioned by Statoil, the Norwegian state-owned oil company, to look at the fossil pollen and marine plankton found within some of the interbasaltic sedimentary layers. The pollen would tell us about the unusual ecology of this strange landscape through the time of opening of the Norwegian Sea, and the plankton would give an age determination as accurate as the geophysical one.

The plankton turned out to be particularly valuable, because there was one species known as a 'marker'. It is thought to have originated 58 million years ago and to have become extinct a million years later. There are many checks of these dates with other kinds of dating, from geophysics, and other fossils, so the figures are quite reliable.

The fossil pollen that we studied from the same samples, however, was less exciting, telling us that there were few species of plants, mainly swamp cypress, pines, a few shrubs like alder, birch and ferns. The ecosystem returned to much the same stable-state vegetation every time there was a quiet break in the volcanicity. I wanted to find out more detail, to check whether these 57–58 million-year-old plants were like any known species. For that we had to look at the leaf fossils that Starkie Gardner collected from Mull a century earlier.

Although Gardner and a few others made some half-hearted attempts to record details of the Mull fossils in the 1880s, no complete account had been published. So I had to go to the museums in London and Stockholm, and I was in for a surprise at both. There were plenty of fossil leaves from Mull in the London collection, labelled with strange names and question marks. Thousands of fossil leaves from elsewhere in the Brito-Arctic Igneous Province, Spitsbergen and eastern Greenland were in the Stockholm Museum, none of them with labels or any hint of botanical identity. There was an air of ghostly mystery hanging over both collections. Yet there was to be a surprise exhibit in each collection.

In London there was a dusty but substantial typescript naming and describing the Mull leaves, produced in the 1930s by Professor Sir Albert Seward (former Vice Chancellor of Cambridge University and Master of Downing College) and W. N. Edwards (curator of the fossils). Also, there were boxes of photographs of the specimens and handwritten notes. Clearly, they were preparing a monograph, which for some reason was not finished.

In Stockholm I found a similarly mysterious document, a bound set of drawings from the best specimens of the shambolic collection. They are exquisite pencil drawings of fossil leaves from Spitsbergen by a very talented artist, but with no labels. There were only two names. One was the artist's, C. Hedelin; the other appeared in the title: 'Nathorst: Tertiary Flora of Spitsbergen'. A. G. Nathorst had been the founder of paleobotany at the Stockholm Museum and professor from 1884 to 1917. Again there had been problems with identifications that had apparently prevented Nathorst from interpreting and publishing the fossil collection.

Svein and I are not experts on angiosperm leaf fossils, so we called in the help of the world's best, Zlatko Kvacek, from Prague. Zlatko and I had been close friends since 1968, when we shared a research exchange fellowship. Twenty-five years on, the three of us worked on the Brito-Arctic fossils for three years and assembled them, together with other Ocean Drilling Program core studies, into an up-to-date account of North Atlantic ecology at the end of the Paleocene. We did it in the knowledge of how flowering plants were changing at the time of their evolution, concepts developed after the deaths of Nathorst, Seward and Edwards. Our evidence showed that the specimens belonged to species and genera quite different from any living today, though most of the Families are familiar.

Nathorst and Seward had not realised that evolution had made names of modern successors untenable for these 55-million-year-old specimens. Their manuscripts had not been finished because they didn't know that the specimens came from extinct species or genera, only the Families being present in the floras we know today. Also, they were unaware that the separation of North America from Europe divides the Families into two separate evolutionary branches evolving on different sides of the Atlantic in increasingly different environments.

The separation of Europe from North America was not complete until 35 million years ago, when exchange of water between the North Atlantic and the Arctic Oceans was first established. Immediately, weather patterns changed because something like the modern Gulf

Stream forced a conveyor of warm water from the Caribbean into the Arctic Ocean, bringing southerly weather with it. There is quite a contrast in the expression of energy between the sudden K–T event and the protracted opening of the North Atlantic, rarely faster than 3 cm a year.

But these Early Tertiary ocean currents were quite different from today's. We don't begin to understand the detailed differences in the climatic patterns, though some of the animal and plant migrations are being recorded in databases for the first time. As usual amidst all this environmental change there was evolution continuing inside the cells as new DNA sequences, waiting for a chance to show the structural consequences when things outside the cells permitted.

But the slow environmental changes through the Tertiary ensured that evolutionary change was also uneventful in comparison to the events following a mass extinction. The warm temperatures up to the end of the Eocene ensured more species and more genera within the Family groups that had mostly been established by then. Evolution was active at this small scale, with no extinctions or origins of large groups. Similarly, after that peak the cyclic cooling encouraged some new species and fewer new genera to originate or become extinct. The more usual response to environmental changes for animals and plants was migration.

Changes in the oceans

As life on land was getting hotter, in the sea things were also changing. The K–T impact may have set off massive volcanic activity in what is now the Indian subcontinent. The resulting eruptions created layers of basalt several kilometres thick, called the Deccan Traps. These major traumas resulted in fires on land from molten lava, followed by volcanic outpourings over an area of more than half a million square kilometres in the north-west quarter of India. At the time, India was an island, moving north from the old Gondwanaland, eventually to crash against Asia, a crash that produced the Himalayas. The Deccan volcanicity caused acidity in the world's oceans, and the amount of oxygen dissolved in the sea fell very low. Following so quickly on the K–T pollution this second outpouring of toxic matter was not good for marine life. The two setbacks to the world's oceans prolonged the time of recuperation needed by the molluscs and vertebrate fish. In the oceans, the Paleocene became a time for slow and steady restocking of the number of new species and genera. Then, about 55 million years ago when the new episode of volcanics began in the North Atlantic, the oceans experienced a different kind of major environmental change.

The temperature of the deep water in the world's oceans suddenly rose, causing the extinction of many of the peculiar species that lived there. Conversely, at the surface, species of plankton bloomed and reached very high numbers for many of the new species. The contrast shows that different organisms can respond in very different ways to the same change in environment.

The causes of the sudden rise in water temperature are not known, but there's no shortage of theories and speculation. The strongest contender is that the opening of the North Atlantic Ocean caused an outpouring of energy from the submerged mid-Atlantic ridge, immediately warming the deepest water. Hot rock from beneath the crust convects upwards to the sea floor right along the ridge between the North American and European plates. At the same time, eruptions on land heated the shallower water by the shores. Changes in ocean circulation resulted from these temperature rises.

Fifty million years ago, the levels and currents of ocean waters in north-west Europe were changing unusually fast, mainly because of the northern opening of the mid-Atlantic ridge zipper. The forces of continental drift caused the split east of the American tectonic plate to leave the British Isles on the European side, though there was still a land bridge from France, through Britain and the Faeroes to Greenland. Sea level was rising and falling every several millions of years, forming sediment such as London Clay at times of marine highs, and coarser sands when the sea regressed. The flora and fauna responded to these regular fluctuations and show up in the fossil record, which help us deduce the environmental changes.

Today, world marine life is under threat from overfishing and pollution; naïve fish-farming practices have led to dramatic loss in populations. In turn there are effects on the phytoplankton which are crucial in the food chain of all these creatures. It is a curious replay of some of the changes that dominated the land and sea at the end of the Paleocene and set the scene for another 20 million years of high temperatures on our planet. Between 55 and 35 million years ago, Earth saw the peak in diversity of very many animal and plant groups on the land and in the sea. They were relatively tranquil times when the planet was largely at peace with itself, content with the steady rhythms of oscillating temperature, sea level and other environmental changes.

Temperatures had been rising steadily since the dinosaur extinction event 20 million years before and flora and fauna flourished to reach a peak in species number. Carbon dioxide concentration in the atmosphere was much higher than today, and without polar ice the sea level was higher as well. Every few tens of thousands of years the river estuaries leading into the North Sea would flood to lay more sediment: the river Thames deposited London Clay from the west, the Seine left deposits from the south around Paris, while the Rhine formed rich sediments in Westphalia. These three 'rivers' were not the half-kilometre-wide trickles we know now. Man-made embankments give a deceptive idea of the extent of natural floodplains.

The rivers were vast expanses of floodplain, up to hundreds of kilometres across, comprising habitats that ranged from waterlogged marsh to quite shallow water and streaming torrents. The proportion of salt water and fresh water varied according to the different environmental pictures. One of the best permanent exposures of one kind of the resulting sediments, London Clay, is on the northern shore of the Isle of Sheppey, in Kent. On the beach at low tide you can find the fossilised remains of sharks' teeth, tropical mangrove plants and tropical rainforest lianas.

The sea level started to rise and fall through these Eocene times 50 million years ago, and the oscillations were the first signs of a change away from the smooth environmental stability since the beginning of the Tertiary. Temperature stopped rising continuously and also began to alternate between highs and lows, rising to a maximum in Europe about 40 million years ago. The low ground between London, Paris and Bonn was at its hottest and tropical rainforest covered much of southern England, Germany and France. As the sea level rose the land bridge separating the London–Paris basin from the Rhine basin became covered with water. Then the sea level fell for another few hundred thousand years to make firm land connections. The land bridge between the British islands and the European mainland rose and fell above these waves. The phenomenon formed cycles that were repeated several times through the 10 million years of tropical heat. Then the climatic oscillations show that it slowly began to get colder.

These ideas of climate change in the Eocene of north-west Europe were confirmed in the 1980s by the geological exploration funded by the North Sea oil industry. Thousands of boreholes cored from shallow-water rigs through the clay offshore show cyclical changes in the diversity and concentration of many different kinds of biology. Mollusc and brachiopod shells varied, plankton gave reliable correlations to the same changes, as did pollen from the plants on nearby landmasses. The times of smoothly increasing temperatures were over. At the end of the Eocene, 35 million years ago, cycles of steady global cooling began to show up in the monitoring of climate change.

Out of the ocean rises Atlantis

One day in 1995 another of those small personal incidents happened to lead me into another area of human fascination, into the lost world of Atlantis. A little package was delivered to my office from Svein Manum, my colleague in Oslo. It contained small plastic bags, each with a few grams of sediment that came from rock cored 350 metres below the sea floor, itself 2 kilometres beneath the sea surface. The source was the northernmost North Atlantic between Spitsbergen and Greenland, about 80 degrees north. No sampling had been done there before because icebergs had made it too dangerous for deep-sea drilling, but now a new ship had equipment to make it possible. The core had been drilled in August 1994 by the Ocean Drilling Program's ship *JOIDES Resolution*, and some of the shales that Svein and his co-workers were studying for plankton also contained abundant fossil pollen. In the knowledge of the vegetation gained from the pollen, and other evidence from the cores, we should be able to tell something about the age of the deposit and the high-latitude environment at the time.

The fifty samples I received came from about every 4 metres in the lower half of the hole. It turned out that the lowest, 350 metres below the sea floor, held sediment laid down 33 million years ago. The youngest, 150 metres from the top of the hole, was just 10 million years old. Evidence from geologists and meteorologists confirms the same story from these sediments. Thirty-five million years ago the new Atlantic, reaching from Greenland and Spitsbergen to northern Norway, had warm ocean waves without storms or strong currents. Slowly, the water halfway between the two continents became shallower with gyrating currents, and the sea floor rose up above the waves. As continental drift widened the ocean, and Greenland separated from Norway, the new island grew to be about 15km wide and 500km long. It had no uplands, just low-lying swampy ground. Remote from any other continental mass of land or continental shelf, growing from the mid-Atlantic ridge, this nameless land was a true continent.

The fossil pollen that I examined from those samples in plastic bags shows a dark conifer forest growing right through the 23-million-year sequence of my samples. There were very few flowering plants, just a few growing in clearings and at the edge of the forest. Instead, there were just dark evergreen conifers and ferns. This was a strange place, unlike anywhere today. There was no Gulf Stream, no wind, trees growing close together, silence. Being so close to the north, the sun was low in the sky, so there were long bright summers and three months of continuous winter darkness. Yet it was always warm and temperate, with no frosts. There is no evidence that mammals ever got there, but if so they would have had a unique life dominated by this strange dark forest. It was dark in the summer from the shade of the dense conifers, and dark in the winter because there was no sun. On the edge of the island there were salt marshes and muddy beaches.

The eerie place stayed like this for 23 million years without environmental crises of change in the vegetation cover. The cyclical cooling in climate, detected from evidence further south on the European continent, had no major effect on these sturdy and hardy dark forests. Twelve million years ago, moved by the complex submarine geology of the mid-Atlantic ridge, the tiny isolated continent was drawn back down under the waves. Now it survives as the Hovgaard Ridge 2km below sea level, midway between Spitsbergen and Greenland.

According to the Greek philosopher Plato, the lost continent Atlantis was devastated by floods and earthquakes, and has not been found since. It's a story that fits with the Hovgaard ridge continental exposure, except that was much earlier than 9600 BC and there was not a city. Plato's myth attracts attention from geographers, historians and archeologists who all seriously try to find scientific evidence for the existence of the missing continent. More than a hundred geographical locations have been proposed, mostly around the Mediterranean, the Atlantic and the Americas, but none are accepted. Curiously the myth is taken more and more seriously, as though Atlantis did exist. The fantasy of the myth is destroyed by the cold scientific evidence, but I'm sure that Plato's story will survive.

These were the relatively docile times of the Oligocene, 35 to 23 million years ago. There was high but decreasing greenhouse warmth and many groups of organisms reached maximum diversity. Without ice at the poles, Antarctica was a well-forested temperate island with one long summer and winter, alternating those seasons with the temperate waters of the Arctic Ocean. Between these extremes, the climate stayed hot towards the Equator, which had a climate similar to today's. As the cooling began very slowly elsewhere, the small range of ecosystems formed by climate began to increase. So the planet was covered with huge tracts of tropical and subtropical forest systems, desert and warm-temperate to temperate ecosystems, with new environments developing at the cooling poles.

During the Oligocene there was a biological boom. The range of mammal species and Families reached a peak on the largest continents, and the seas were much warmer, with more Families of fish and plankton than before or since. Of course there is change hidden in the apparently quiescent layers of sediment – oscillations of temperature, rainfall and all other weather. Otherwise, the warm world of the Oligocene was one of the most peaceful and stable times in the history of life on the planet. Apparently without major environmental events, there were no large meteorite hits or major continental crashes. The evolutionary machine was running efficiently in a beautifully controlled style, every gear wheel fitting snugly with perfect timing. It was a peak of global success.

Cool rhythms of the Miocene

Gradually, this monotonously running machine needed to adjust to the cooler temperatures. Carbon dioxide from the atmosphere was being used up by reacting chemically with the rocks from new mountain chains like the Himalayas, and by increased photosynthesis; meanwhile the unusual paucity of volcanic activity over tens of millions of years didn't replenish stocks. Other changes in climate were due to things like moving continents forcing new oceanic currents. There was a very slow change to a more variable mix of different biology, environment and climate. The cooling was often interrupted by warm cycles lasting hundreds of thousands of years.

Equatorial climates continued their equable state, allowing steady evolutionary progress for the highly diverse animals and plants evolving there. We know less about the tropical and subtropical systems because erosion has taken most of the remains, and there have been few special-ists able to explore the hostile territory. As usual in the affairs of natural history, it is from the developed world that we have most data. In the Miocene, 23 to 5 million years ago, much of our knowledge comes from browncoal mines in Europe, where many specialists have spent their lives trying to understand the animal and plant remains in the lignites. At the end of the Cold War, many specialists from the Soviet empire were able to publish lots of information and ideas that had been accumulating through the dark years of censorship since the 1930s. Now there is a surfeit of data, biased to the north.

At the beginning of the Miocene, or perhaps it was just before, a simple event, experienced across the northern continents, caused a phenomenon not seen on planet Earth for 250 million years, since before the P–Tr mass-extinction event. Frost. We first detected it from pollen diagrams, because palm pollen suddenly disappeared. Palms had been an important part of floras globally and they will not tolerate sub-zero temperatures. Suddenly, around 25 million years ago, the distinctive palm pollen is absent from northerly localities. At the same time, the few warm-loving relicts from subtropical forests also vanished from the north. The familiar northern mixed conifer-deciduous forest was settling in around the now colder Arctic Ocean.

The strangest observation to be made about these forests must have been the very high proportion of conifers from a Family that is nearly extinct today but grows in a wide range of habitats. Living plants of the Taxodiaceae are relicts restricted to very small regions. The best-known are the *Sequoia* redwoods in California and *Taxodium* swamp cypresses in Florida. In south-east Asia there are other genera of single species with very small numbers of natural individuals hidden in deep forest. The best-known are two forms used widely to decorate road-side verges. One is *Metasequoia*, whose living species was found after the fossil had been described earlier, and then there is *Sciadopitys*, the Japanese umbrella pine, which hates frost.

Throughout the Miocene and the 4-million-year Pliocene period that followed, these genera, and extinct ones like them, played a major role in the ecosystems of the increasingly temperate northern hemi-sphere. From northern Canada down to California, northern Greenland down to the English Channel, and from northern Siberia down to Vietnam, they formed conspicuous elements of the landscape. *Sequoia* in the mountains and hills, umbrella pines in the sheltered valleys, *Taxodium* in the wetlands. Of course, they were all mixed up with other conifers and deciduous flowering plants, but they were often the dominant group.

It follows that through the Miocene, European uplands looked very similar to present-day California and the lowlands more like Florida's swamps. Because the continents were separate the species and some genera were different, but the broad composition of the vegetation was similar. The mammals also shared DNA from the same evolutionary branches and their diversification depended a lot on the kinds of forest they inhabited. For their evolution to proceed, some break was necessary in the monotony of the vast northern *Sequoia* forests and *Taxodium* swamps. It happened in two ways: through increasing variation in the climatic oscillations that began in the Eocene 50 million years ago, and through the later evolution of grassland in America and Asia that was to influence in a spectacular way the way animals feed. The Miocene warm phases may each have lasted up to a million years, alternating with frosty intervals which may even have enabled ice to form at the poles. Eventually came the formation of permanent polar icecaps and an unusually high range of environments between them.

At one of the cycles' peaks, about 12 million years ago, the climate in Europe was unusually warm. The frost-sensitive umbrella pine conifers reached a maximum in numbers and distribution. Up to twenty species of warm-loving tender shrubs and delicate ferns appear in many of the floras that have been described. As well as warmth, high rainfall is necessary for their modern equivalents, and can be expected to have been part of the European environment. One of the most convincing groups of fossils to support this theory is the 12-million-year-old banana fruits discovered a few years ago in Denmark.

The more extreme seasons encouraged many new biological features to adapt to each new situation. One is seasonal deciduousness, as well as the evolution of herbaceous plants with annual life-cycles. Then, around 10 million years ago, grass prairie finally appeared in parts of North America and Asia. Grazing had begun and necks started to bend down to the ground. What was good news for these new mammals on the prairie and open forest was bad news for the browsers that had adapted to reach up to the high leaves of the broadleaf forest.

The flight of the browsers from temperate regions was almost complete at the end of the Miocene, and they took refuge in much smaller numbers and places around the tropics. For more than 20 million years these mammals have been in decline. It's not only necks that distinguish the two groups of mammals, it's the teeth as well. Most grazers don't need to chew large leaves: they don't grind but slice, biting herbs and grass, swallowing whole for slower digestion. The new grazers adapted different teeth. Their sharp incisors had high crowns that can cut the herbs and grass very effectively. These plants are often very different foodstuff from the broad-leaved diet that browsers ate. Recent biochemistry research suggests that that's because the herbs and grasses photosynthesise differently, producing hard crystals as waste products that also toughen the plant's tissue to help its survival. That may have been the result of lower levels of CO_2 in the atmosphere, changing the biochemistry of photosynthesis. Chemical fossils from ancient soils and in fossil teeth from horses suggest that the changes happened between 8 and 6 million years ago, precisely when other evidence shows that the harder new grasses evolved.

But this grand theory is beginning to attract a lot of criticism and may be another oversimplification, needing input from several disciplines. One difficulty is that herbs and grass don't fossilise very well. The pollen is about all we have, and its fossil record shows very little change through the Miocene and Pliocene. It's impossible to distinguish between the different types of herbs and grass. All the evidence shows is an increase in the numbers of individuals in association with heathland and new pioneer forest.

What is clear is that CO_2 concentration in the atmosphere was falling, and the temperatures were lower through the influence of a reduced greenhouse effect. That may have triggered some changes to the biochemistry and physiology of photosynthesis. There is also some evidence from geophysics. Isotopes found in fossil soils and teeth of grazing horses show the same match: broadleaf plants usually occur with browsers' teeth while the grazers are associated more with the tougher grasses. The browsers with their low-crowned teeth were unhappy with the tougher grass and became extinct on the expanding prairies. The connection between differing photosynthesis in broadleaf plants, herbs and CO_2 concentration is not yet clear, so we need more research to clarify the relationships.

As we shall see, there are other specialists who are beginning to challenge the role of atmospheric CO_2 alone in causing global changes in temperature. Instead they see a complex interaction between processes in the northern and southern hemispheres including changes in ocean currents. Taking these uncertainties with the speculation about grass evolution and the biochemical changes in photosynthetic products, there is no wonder that there is confusion about these complex systems. We may have to wait some time to understand how mammals became large and how some of them changed from browsers to grazers.

Towards an icehouse world

When the sun shines on the plants and surfaces inside a greenhouse, some energy is absorbed by them and the rest is reflected upwards off the leaves. But the glass in the building reflects it back inside again, and again, so forming an energy trap that absorbs heat. The greenhouse effect works when high levels of CO_2 and other large molecules in the atmosphere, serving the role of the glass in the greenhouse, raise the temperature of the plants and the surfaces on the planet. Low levels of CO_2 reduce the trapping of these reflections and allow the sun's energy to go back out into space. This is the characteristic of an icehouse world.

One such event started at the end of the Miocene as atmospheric CO_2 levels continued to fall. The cooling climatic oscillations that began in the Eocene continued at one extreme to reach freezing temperatures. This is another theory that relies heavily on the changing concentration of CO_2 in the atmosphere. Maybe the building of the icehouse was more complex. Maybe it had something to do with the north/south interaction I mentioned earlier. But for now the fall of CO_2 below Miocene levels is the favourite explanation of the Earth's cooling over the last 35 million years. Whether such a simple idea can account for the most extreme cooling, over the last 2 or 3 million years and leading to the ice ages, is another matter.

The cause or causes of the ice ages is still unclear – one of those difficult riddles involving changes in systems outside the Earth over more than the last 5 million years. No wonder that we have no confident answers about the cause, or any integrated trends from the disciplines involved – astronomy, physics, biology and meteorology. Is it, perhaps, another self-organised system waiting to be discovered?

Some of the most important early work to try and understand ice ages was done at Cambridge after the Second World War. It was led by Professor Sir Harry Godwin, head of the Botany School and a lover of the flat expanses of East Anglia and its flora. To the north-east of the city are wetlands, mostly fens drained centuries ago to protect fertile agricultural land from high storm tides of the North Sea. Much of East Anglia and the Netherlands has been reclaimed in this way, encouraging scientists like Godwin and van der Hammen, a Dutch pollen specialist, to study pollen diagrams from sections showing changes in ecology. Through the climatic oscillation of one ice age, the ecology passes from glaciation to permafrost, cold to warm forest, and then back again to the next glaciation. The unfrozen part of this sequence gives vegetation and fauna showing a very characteristic set of changes in ecological succession through the cold–warm–cold of an interglacial.

In their day, Godwin and van der Hammen were fêted, and their students continue to work through the details of what happened between all the ice ages. With their colleagues they described and counted pollen from hundreds of interglacial and other sites deposited since the K–T catastrophe. Pictures emerged about the successional ecological changes between glaciation and the millions of years before. Up to the 1970s clear reconstructions of the changes seemed to be emerging, and four or five glaciations were agreed to have covered north European landmasses through the Pleistocene. They put all the little bits of evidence together with thousands of sites and short sections. Yet even the longest time sequence represented only a few thousand years, and often the precise date of the age of the sediment was unsure. Even Godwin's group admitted it was all a bit fragmented.

Then came a shock. Many of these ideas were toppled when continuous deep-sea cores became available. Put side by side, they are able to show the full 1.6-million-year time frame of the ice ages. The cores were well dated by many techniques, full of small fossils to reconstruct the environment in which the sediments were formed, and they came from continuous non-stop deposition, right through ice ages and interglacials. One thing became clear quickly: the climatic changes through the last 2 million years were much more complex than had been thought, and there were many more oscillations than had been detected from the land-based evidence. Instead of four or five major ice ages, it looks as though there have been at least ten through the last one million years.

One of Godwin's successors in Cambridge was Nick Shackleton, a physicist interested in calculating the age of marine microplankton from the Ocean Drilling Program sequences. He measured the proportion of different isotopes of oxygen preserved over time in the shells of particular plankton species and calculated the temperature changes very accurately. Together with data from other specialists, his work showed that the complete ice age cycles over the last half million years had essentially the same frequency, one cycle every 100,000 years or so (see figure 4.3). The regularity of the changes made the group think that they were caused by changes in the shape of the Earth's orbit around the sun.

This idea that astronomical changes between the Earth and the sun caused the ice ages was not new. In 1875 the Scottish physicist James Croll suggested that ice ages were caused by changes in the radiation from the sun with a frequency of 100,000 years, in the tilt of the Earth every 40,000 years, and in the planet's wobble every 20,000 years. In the 1930s Milutin Milankovitch, an engineer from Serbia, detected a

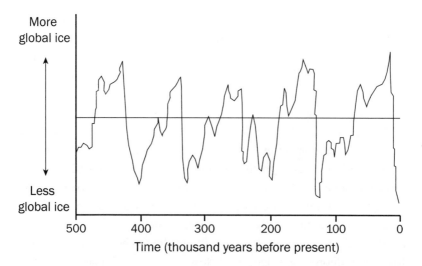

Fig. 4.3 Curve representing changing temperatures through the last half million years. The time from one peak to another represents a full glacial cycle, due to the changes in the shape of the Earth's orbit around the sun. The data are from oxygen isotope recordings in shells of marine microfossils sampled from the Vostok ice core in Antarctica. (after Schrag 2000)

range of orbital variations on the Earth, roughly at 20-, 70-, 100- and maybe 400-thousand-year cycles. Now we suspect that other planets in our solar system show comparable changes and may influence the oscillations here. But the main weaknesses of Croll and Milankovitch's ideas are shown nearer to home. How can the small changes in the oscillations cause such large changes in temperature? Why do ice ages occur in both hemispheres simultaneously when solar radiation has an opposite effect on the north to the south, causing one to be winter when the other is summer?

Another theory to explain the origin of the ice ages is different from Milankovitch's but still relies on the sun. Roughly every 100,000 years the plane of orbit of the Earth around the sun passes through a cycle, though the shape of the orbit stays the same. The Earth could pick up dust from clouds in another orbit, with this regularity and with the ability to reduce radiation from the sun. In the next section I describe some evidence for the possible effects of such extraterrestrial dust on our planet, but it's unlikely there is enough to stimulate enough cooling to cause the ice ages.

Through the last four or five years there has been a lot of evidence published to support a third theory to explain the ice ages. Between 1.5 and 0.6 million years ago the climatic oscillations changed to cover a wider temperature range over a longer time interval. Before 1.5 million years ago the temperature range was ten times less than in the present ice age cycle, and the oscillations occurred much more frequently. Then, it was small and often, now it is large and slow. The new explanation is that 1.5 million years ago, heat-flow across the Equator started to increase, changing Earth's orbit, especially its closest approach to the sun, and triggering the longer, more extreme glacial cycles.

Some idea of this north/south hemisphere exchange comes from another very new theory of modern climate change. A few years ago the concept of a '*North Atlantic Conveyor*' was proposed, described as bringing more warm water north from the Caribbean to Greenland. It's a super Gulf Stream, melting a lot of Arctic ice and increasing the amount of less dense freshwater. The changing ocean currents may cause very sudden cooling in the north. The theory is that this northern hemisphere process balances effects caused in the south by the El Niño current. It means that a vast amount of warm water in the Pacific Ocean is a dominant force affecting climate. The interactions between north and south may have influenced polar temperatures for more than a million years.

From millions of years to tens

Some of the most spectacular and informative means to measure climate change very accurately are two ice cores, *Vostok* and *GRIP*, taken from bore holes in Antarctica and Greenland. The refrigerated cores from these holes are more than 3,500 metres long and the telegraph-pole-like sections contain samples of the actual precipitation of rain deposited over the last half million years, together with air bubbles, fossil dirt, dust and other particles carried by the ancient atmosphere. Modern engineering and dating techniques mean that the age of each layer can be fixed to within ten years near the top of the cores and a few hundred further down. These curious sources of polar evidence inevitably attract comparison of changes in the north and south. Just as there seems to be some connection between the Gulf Stream in the North Atlantic and El Niño in the South Pacific, so results from the poles show the operation of a north/south seesaw effect. They show that the southern hemisphere got warmer a few thousand years before the north and that these cycles lasted through the whole of the last glacial period.

The huge amount of data coming from studies of these ice cores in the late 1990s is only now being published, and a lot is available on the internet. Sorting it all out will be a long job. For example, we don't know whether the GRIP core contains evidence of yet another climatic event that occurred relatively recently, in AD 540. It's one of a growing number of changes that have been detected in the biological record that can be correlated with historical events, in this case King Arthur's death, the plague and the end of the Roman empire in Britain. From China, west through Europe to North America, narrow tree rings have been dated to that year. These signals of low growth mean that extremely unusual cold distinguished the winters around that time. Disease and social unrest caused havoc and hard times for the struggling people of the northern hemisphere, and the so-called Dark Ages began. There was intellectual and social quiescence which was to last for many hundreds of years. We can still find evidence for those quiet times from the fossil record that was left behind in the ground, in the peat, the lake sediments and the soil. Is it in the ice cores as well?

Another example of a cold snap, the Younger *Dryas* Event, comes from studies of a period 13,000 to 11,200 years ago when the rising temperatures after the full ice age slowed down. The retreating glaciers in the northern hemisphere suddenly got colder and grew forward again. We first knew about this from an increase in fossil leaves and pollen dispersed by a small shrivelled buttercup-like weed, *Dryas octopetala*, which still grows well in the steppes of Arctic landscapes. It is thought to have been caused by the Gulf Stream suddenly stopping, so that the average annual temperatures in England were as low as $-5°$ rather than the $+11°$ of today. Why the Gulf Stream should have stopped then is anyone's guess. If Arctic snow and ice melt enough freshwater, pushing the denser salt water from the surface of the sea, the Atlantic Ocean circulation might suddenly flip into a cold mode. Some specialists have suggested that this happened to cause the Younger *Dryas* Event. Others are suggesting that it might happen tomorrow.

It appears that there have been several cold snaps like this since the last full ice age. The *Dryas* one was longest, and used to be thought to have been restricted to Europe and North America, around the North Atlantic Ocean. But new evidence out of the Amazon basin shows much-reduced discharge of sediment out of the delta at the same time, as would be expected during a cold period, even in the tropics. We simply don't have enough information from around the world, and from different disciplines, to be able to link the oscillations to one another or to other particular causes. Furthermore, might these Atlantic processes be related to El Niño in the Pacific?

Between the cold snaps came the balmy days of warmth, sudden and short high peaks in the curves of climate change. In the north, between AD 900 and 1300 cattle were farmed in Greenland and the French tried to embargo English wine. Nevertheless the good weather ended as quickly as it had begun. In five years famine was to replace the harvests, and the Black Death of the 1340s saw an end to warmth as well as happiness.

More recently than the *Dryas*, in the early 1600s temperate climates around the North Atlantic once again became much colder. This snap is called the 'Little Ice Age' and lasted until around 1850. In North America the cooler climate affected the vegetation and encouraged a species of grass to which the buffalo, or bison, were especially partial, but horses were not. The buffalo increased, the horses declined and a new balance between the fauna and flora was established. We know partly from oil paintings that on the other side of the Atlantic the river Thames froze in London twenty-three times between 1620 and 1815.

That these changes focus on the North Atlantic suggests that the causation is regional and therefore within the planet, rather than external. Later, in chapter 6, I describe how rife the speculation is about forthcoming changes in the Gulf Stream, and how it seems to be caused by our modern lifestyle. But is this *Dryas* and 'Little Ice Age' evidence saying that the same thing happened before humans started to change so much of the environment? Were the causes coming from humans or nature? The same question is asked of the next stage of the North American buffalo story. Just as the climate was warming, in the 1840s, historical records tell us that Red Indians began trading buffalo hide. One view is that hundreds of thousands of buffalo perished in nature's cold. Another ascribes the slaughter to men, hunting in specialised groups and coming from Asia.

This mixture of human history with ecology, climate change, human culture and fable is hard to understand as a single system. Which explanation of the fall of the buffalo are we to accept? Were extraterrestrial forces responsible for stimulating some of the changes? Were the changes caused by aggressive man, extraterrestrial clouds of dust, sunspots or something unknown from within or without? Perhaps it's due to all these forces, in different places at different times. Patterns of evidence tie together different ecological balances, stable climaxes upset by human and climatic interventions. After each event the system recovers, often quite quickly in a few human generations: only then to be upset again by more environmental changes. That seems to be the way for so much on our Earth.

One way through these mysteries is to take the most reliable evidence of past environmental and evolutionary change and see if it follows any clear pattern or mathematical model. As well as demonstrating power laws, implicit in the concept of self-organised systems, does the huge complexity of the enormous system of planet Earth over the last 65 million years display patterns or trends? Is there evidence that major groups of animal and plant Families follow clear rules in their origin, expansion, peak diversification and eventual extinction?

In 1999, twelve years after Per Bak's famous paper in *Physical Review Letters* about concepts of the sand pile, the same journal published an idea that applied Bak's treatment directly to the species-and-area relationship. The aim was to see whether these relationships within an ecosystem are self-controlled: whether the changes in geographical distribution follow the patterns of power laws and pink noise. The article was written by Jon Pelletier, a geologist at Cal Tech. Sure enough his model does show that the number of species depends on the area, and it gives the familiar power law and pink noise. This time the theme is biogeography: changes of animal and plant distribution through thousands and millions of years seem to show the by now familiar self-organised pattern. More has to be done to understand the extent of this feature, especially to find out what role environmental change plays in the relationships. These new studies may mean that whole ecosystems – animals, plants, ecology, climate – can behave as self-organised systems. When they are in a stable state, with only minor evolutionary and ecological changes, they will control their own survival just as the sand pile does.

5

What's in a Name?

After our breakthrough confirming that evolution follows an exponential path, with proof from features the experts call power laws and pink noise, my research group was feeling on tenterhooks about the hidden potentials of our techniques. We had come to realise that we were the first to mine the database in this way and so we might expect more prizes from the data. So we moved on quickly to search for other patterns that might reflect on issues of evolutionary biology. An obvious interest was to find trends in the groups of animals and plants that have lived on this planet. Do they have something in common in the way they originated, reached peak diversity and eventually became extinct?

It was then that our diverse backgrounds showed up in an unexpectedly positive light. I am the only one of the group trained as a biologist, so to me words like 'species', 'order' and 'phylum' are everyday terms which are taken for granted. So when I asked my colleagues 'to find trends in the groups of animals and plants that have lived on this planet' the request seemed very straightforward to me. But not so to a mathematician, a computer scientist or a statistician. They wanted to know what 'species' or 'phylum' mean and how they differ from one another. 'Only then can we begin to compare one with another, confident that we are comparing like with like. Otherwise the results will be chaotic and meaningless.' I was in trouble. I couldn't help them understand. The books I gave them made matters worse because each sang a different tune.

The building blocks of God's nature

Aristotle was the first scientist to write at length about plants and animals, around 350 BC. He was thinking about different groups of organisms and linking his ideas into many other disciplines, some practical like meteorology, some more theoretical like logic and metaphysics. In his lecture notes 'living things' are divided into two different categories: 'plants' and 'creatures'. Then the 'creatures' are subdivided into 'animals' and 'humans', and so on. This was at the opposite end of the spectrum of knowledge from where we are today. He was at the beginning, breaking new knowledge of the world into bits to work on separately, whether they are plants, weather systems or ideas for a literary drama. Disciplines were being born, and they have stuck with us ever since as separate intellectual entities. Some of them, like physick and alchemy, are largely ignored today, but most classifications of knowledge still follow Aristotle's early categories. They also included natural history, astronomy, analytics, ethics, poetics and metaphysics.

Aristotle was also interested in linking these parts together into hierarchies, forming trees of knowledge. So the different divisions of animals and plants start with the simplest structures and end with the most complex, humans. Each variant of form fits into a different box, all neatly positioned in order of ascending complexity. Each box corresponds to a different living form and assigns an identity to each division.

Today we're busy putting some of Aristotle's divisions together again, both within biology and between disciplines. But we are doing it in a very different way, especially when tackling broad problems like those of biology and the environment. We have less need to see the world so neatly and can live more confidently with both chaos and complexity. Some very timely help with this is coming from the internet. Network structures reflected in search engines like *Alta Vista* help us find data from different subjects that are variously relevant to the theme. We can link things together that have never been connected before. Searching for 'Aristotle' gets thousands of different accounts and interpretations of the man and his ideas, and often obscure applications and commercial marketing.

Aristotelian logic had the task of discovering 'pure knowledge', now known as 'science', and described the true nature of things, their inner reality or 'matter'. 'Development is realised when the last form is reached. This is called *matter*, by which is meant all those conditions which make possible the passage of successive forms. To function thus, matter must remain unchangeable.' Aristotle thought nature was made up of fixed and concise parts which would eventually be described as true order, without chaos or clouds of subjective interpretation. That's why biologists have come to think of organisms with the same natural form as members of the same species. This idea of distinct entities with clear definition was taken up much later by Linnaeus.

Carl Linnaeus (1707–1778) worked from the ideas of Aristotle and fitted them into the Christian values of the times. Aristotle's idea of *matter*, comprising small building blocks, was attractive because it accommodated a truth, a religious goal – the belief in God's creation, as set out in Genesis – which also applied easily to the latest knowledge about animals and plants. Inevitably the fixed blocks became the basis of how organisms were to be named.

In the early eighteen century, Sweden was far away from the new ideas of the Enlightenment slowly developing in southern Europe. Throughout Carl's long life his country was prevented from developing new ideas by a series of political misfortunes. Sweden's young king, Charles XII, had inherited the throne at an unfortunate time, to oversee the fall of his Baltic Empire. This was marked especially at the battle of Poltava in 1709, when Russia took the eastern territories, leaving Charles to reign over a depressed society. It was not to be for another half-century that new ideas from southern Europe would challenge the long-held values. Sweden was denied the influence from the thinking being stimulated by the new sciences, especially from their southern ally, France. Linnaeus was born two years before the battle of Poltava, so he was brought up strictly along the rigid God-fearing lines of these times.

Linnaeus believed that an object could be understood in two ways: the aspect of matter known as its 'comprehension' or the sum of its qualities; and the aspect known as 'extension' or the range of its application. These two concepts are in inverse relation to one another: the greater the comprehension, the narrower the extension, and vice versa. So, if we increase the comprehension of the concept 'animal' by adding another element, for instance fur, the extension of the concept will decrease because it is no longer applicable to all animals, only to mammals. From this Aristotelian reasoning Linnaeus established the 'species' concept so as to name the fixed truth, the irreducible basic unit of matter in nature's order, as Aristotle had required.

Although at one stage Linnaeus listed genera together under 'ORDO', he had no intention of making a hierarchical classification. Beyond *sapiens* there was *Homo*, and the higher names were not invented: Hominidae, Catarrhini, Primates, Eutheria, Mammalia. There was to be nothing but the first two of these ranks, species and genera, hanging together as the world of living organisms. Although this way of thinking left a lot of open questions, it followed the main value of the time: nature was fixed since Genesis and ordered by God. As if by accident the species also offered a new tool, a formal name in a hierarchy, for naturalists to describe plants and animals in the new discipline of biology.

So it is 'species' that came to be seen as the Aristotelian essence, the basic unit that cannot be changed. The genus is at a higher elevation, meaning it is applied to a broader group: Aristotle's field of application or extension. Higher taxa in the hierarchy – Family, Order, Class, Phylum – came after Linnaeus and are more flexible. For example, our own species, *Homo sapiens*, is nested in *Homo* then Hominidae, which the science of taxonomy calls a genus and a Family respectively. But then there are the ranks named Catarrhini, Primates and Eutheria, all around the more uncertain levels of Order and Class. Different people use or ignore these parts of the hierarchy according to their needs. Some experts agree that the name Mammalia is at the rank of Phylum while others do not. The neat organisation of Aristotle's unchangeable pure knowledge is beginning to show some confusion to allow different interpretations by different people.

Certainly there is room for confusion, or at any rate variation, in the biological content of the different rankings. Nowadays, the system allows different specialists to organise the branches differently, because different groups of organisms have different things in common and can fit on different branches of the evolutionary tree, according to how important the common features are thought to be. But is this system, devised 250 years ago, based on a style of thinking developed two thousand years before then, the best way of organising the names of plants and animals in the twenty-first century? It leads to an innate confusion between classification and evolution: whether there is a difference between the stratified hierarchy of classification and the branches of an evolutionary tree.

Unfortunately many non-scientists still expect science to give the right answer to questions of nature. They want to know if it is safe or unsafe to eat beef, or to sow genetically modified seed, and if the scientists don't know the answer, the government will give them three years to find one. So many people expect there to be a truth, and government, journalists and teachers reassure them that that is the way knowledge is. I fear that this is also the way some still understand the long sweep of natural history.

Linnaeus died in 1778, too early for him to be influenced by the new Enlightenment. The strong cultural links between Sweden and France were renewed soon afterwards, and so Linnaeus' species concept was adopted. With their enlightened attitudes to the new biology the French biologists produced the first encyclopedias of the plant and animal kingdoms. These were exciting times intellectually, especially in France, where the new ways of thinking largely originated. Georges Buffon (1707–1788) began the movement with a grand summarising work, *Histoire Naturelle*. This led to the emergence of four grand contemporaries, with different ideas stimulating new attitudes to the subject. The oldest was the rebel against Linnaean species, Jean Lamarck (1744–1829), with his well-known but now discredited theory of acquired characteristics somehow changing a named species. More conventionally, Antoine Jussieu (1748–1836), professor at the Jardin des Plantes, wrote a classification of animals and plants, *Genera Plantarum*; Georges Cuvier (1769–1832) contributed to the classification and anatomy of animals; and Louis Agassiz (1807–1873) was one of the first to specialise on a single group, the fishes.

Large quantities of new data came from these and other first methodical descriptions of animals and plants, and arguments started about which characteristics specialists were to use to sort out all these different organisms. There were new questions about whether the first subdivision of the mammals should be based on the presence of a placenta or a pouch rather than the number of toes. The specialists soon realised it was not going to be easy. It's uncanny how the situation being recognised then in Paris is so like what we face now with the internet, the need to sort out large new sets of data. Just as the concept of Aristotle's essence had worked well for thinkers for over two thousand years, so Linnaeus' 'species' has done well for the last two hundred. But with so much new data from different disciplines we need, today, to move on once again. Scientists will continue to study species and their origins; but with so much new data from different disciplines, we can also explore the broader systems and see some of the patterns they create.

Remember that these pioneers were ignorant of many basic things in evolutionary biology we now take for granted: genes, mutations, chromosomes, the age of the Earth, continental drift, let alone mass-extinction events. Against all these ignorances, Darwin started off with a broad enough mind to be sceptical of the less scientific opposition. Inspired by his observations and thoughts on the *Beagle*, he eventually presented the idea of natural selection.

There is no stated 'Theory of Natural Selection', as biologists don't have theories the way physicists do. Instead, Darwin expressed a grand idea as lots of observations and often anecdotal interpretations, about which much has been written over the last century. We know now that evolution is a very complex cellular process influenced by organisms' responses to the environment, and that it is controlled by the genetic processes of recombination and mutation in the cell.

The relationships between these cellular processes and the organ in which they grow, and between the organisms and their community, are another very complex set of interactions which we are only just beginning to consider together as the same system. We are not yet able to distinguish an actual evolutionary event in the cell at the molecular level, and can only assume that they have taken place from our observations of groups of organisms. This is a pretty large jump to take, but the latest research is making links between DNA sequences and structural characters. It offers a new dimension in understanding evolution.

I must admit to having had trouble for years understanding Darwin's concept of 'natural selection'. What is natural and what is it that's being selected? Darwin's *Origin of Species* doesn't help much, and nor did the nineteenth-century geneticist Gregor Mendel. Once again, there was ignorance about the two concepts: 'species' and 'genes'. Nevertheless there was a breakthrough in understanding the processes, in this case the recombination of genetic material during sexual reproduction. Knowledge of all these things was changing so quickly through the twentieth century that no stable idea of 'natural selection' emerged.

My own trouble was not being able to see a connection between DNA and ecology. Darwin and most biologists of the twentieth century made little connection between them. Still, the two fields tend to be studied by very different sorts of people, using very different kinds of evidence, never sharing their writing in the same books and journals. Then, one day, it clicked. Sex and environments are intertwined. Genes don't copulate, individuals do, and they do it in special places: most humans have beds, some birds have special trees, fish have special currents. The resulting sexual act mixes genes, causing them to recombine and create new offspring. It is the organisms that need special environments to recombine the genes.

Darwin leaves town

Darwin set off for South America on the *Beagle* late in 1831, and spent much of the winter journey suffering from sea-sickness and loneliness. At the same time, in the cosiness of Paris society a Swiss doctor called Louis Agassiz, interested in fish, sought the help of the leading French naturalist Georges Cuvier. The great man was impressed, especially by Agassiz's Latin descriptions in *The Fishes of Brazil* and by his deep faith in God. Agassiz went on to specialise in Alpine glaciations as well as ichthyology.

His interest in glaciers was due to more than his love for the Swiss mountains. While he was in Paris he became involved with many of the intellectuals of the day, and strongly defended conventional biblical wisdom and creationism in particular. That meant that he believed in the Flood, and that he had a responsibility to find evidence to prove the ideas scientifically. But this is not the way that science works. You don't answer a question and then set out to prove it. You test an idea and then look at evidence for and against it. Darwin was doing that from the *Beagle* while Agassiz was examining glaciers.

Searching for more evidence of the Flood, Louis decided to emigrate to the United States. There, his ability to concentrate on more than one discipline at once and his fine reputation from home led to his meteoric rise to the chair of Natural History at Harvard. At the same time he helped his daughters teach at a young ladies' college, and became a founding member of the National Academy of Sciences, sinecures from his intense research on fossil fish and his ever-serious religious beliefs.

One of his monographs included the so-called 'spindle diagram' (see figure 5.1) which has become famous as the first attempt to show diversity changing through time. The thickness of the spindle represents the number of species and the vertical scale is the geological age or the rocks' strata. Most cunning of all is the positioning of the spindles in relation to each other. The diagram implies that the closer the spindles are together the closer they are related. For 1833 it was the

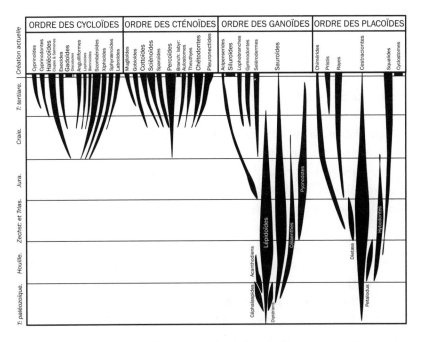

Fig. 5.1 Spindle diagram from Louis Agassiz 1833 *Recherches sur les poissons
fossiles* 171, Neuchatel: Petitpierre. Unlike Darwin's evolutionary tree
(figure 3.4) these lineages are not connected branches. Most of the shapes
reach maximum width (highest diversification) early in their range. The
best examples are the large spindles near the middle of the drawing,
'Lepidoides' and 'Sauroides'. Their shape is the basis for the bell-curve
model, expressed by the equation on page 149.

most far-sighted contribution to visualising evolution, but for one simple
reason Agassiz missed the recognition that usually goes with genius.
Right up to his death he insisted that the creationist outlook on life's
grandeur was correct and that Darwin's ideas on natural selection were
wrong.

The pull of the Roman Catholic Church in Paris just after the fall
of Napoleon was immense. So Cuvier's theory of catastrophism, the
sudden and violent changes that led to differing rock formations,

claimed widespread support and surely influenced Agassiz. Adherents to this theory believed that occasional catastrophic events, such as the biblical Flood, caused extinctions, followed by new creations of life. In this way, the fixed species of Linnaeus could become replaced within the creationist world of Paris's natural historians. It is now clear just how close to showing evolutionary relationships Agassiz had been. His spindle diagram hints at so many of the things involved in the evolution relationships of a large group, in this case fish.

The closeness of some Family spindle shapes suggests a particular closeness in their relationships. But the groups never touch and Agassiz was adamant to keep them separate. Any connection or hint of descent would lead to the recognition of a single evolutionary tree. The diagram had got halfway to that conclusion in another of its innovative features, the shape of the spindles or the shape of change in all biological Family diversity through geological time. This shape of Agassiz's spindles has inspired thousands of teachers to portray evolution diagrammatically. The spindle drawings, now variously linking together, are common icons of evolution in modern textbooks.

Agassiz's diagram was published less than twenty years after the defeat of Napoleon at Waterloo. Cuvier and his colleagues were making daily advances in observations of new morphologies of animals and plants, their classification and their relationships. They were largely ignored across the Channel in England, where even Linnaeus had received little attention. One pioneer who was blissfully unaware of the conflicts between those who give names to organisms and those who classify them was Darwin himself.

The Origin of Species contains only one figure (see figure 3.4), which does link the spindles together. Today we call the figure an evolutionary tree. We also worry more than Darwin did about the identity of each intermediate in the figure. Darwin called his upper-case A–L 'species', and his lower-case a_1–z_{10} 'varieties'.

One thing is clear, Darwin's species are something quite different from those conceived by Linnaeus: Darwin's evolve while Linnaeus's are fixed names on a list. The difference is an issue that Darwin did not consider and which we have started to take seriously only very recently. Another difficulty, easy to see now but hard then, is the unit of time between each of the horizontal lines. Darwin wrote that these were generations of each variety: 'a thousand generations; but it would have been better if each had represented ten thousand generations.' Clearly he was unsure of the timescales.

These interactions were first put into writing by Darwin in a private 189-page essay that he finished in 1844. Troubled by the consequences of lighting this dynamite in front of the theologically-based scientific establishment, he kept his manifesto strictly to himself and a small number of trusted scientists such as Charles Lyell at Cambridge and Asa Gray at Harvard. A few years later they became concerned that another naturalist, Alfred Russel Wallace, was close to publishing similar ideas of a 'natural progress by evolution'. So it was that extracts from Darwin's 1844 essay were read to about thirty fellows of the Linnean Society at one of their ordinary meetings in the summer of 1858. Wallace's manuscript was also read at the same meeting.

There was no fuss, no sense of occasion. Neither Darwin nor Wallace was even present, and the lectures were read by the Secretary. The agenda had been hastily patched together and no one in the audience knew what was coming. Some of those who were present grumbled that the topic that evening had been too theoretical and hoped for more descriptions of new species next time. But Darwin had gone public to win the prize of recognition as the author, whatever the consequences for theology and science. He had no alternative but to reinforce the lecture with *The Origin of Species*, which he finished writing a year later. It was published to storms of invective and praise that took more than fifty years to settle down.

The chair of Natural History at Harvard seems to attract professors at the centre of the debate between creationism and biological evolution. Louis Agassiz believed profoundly in the former. A more recent incumbent of the chair is Steven Jay Gould, who believes profoundly in the latter. The debate will go on.

Against species

In the first half of the twentieth century mainstream biologists described structures and mechanisms. They built a hierarchy of names for the ranges of animals and plants in the evolutionary tree. At the same time genetics and ecology started to develop at an increasing pace. The early ecologists last century didn't take evolution on board at all. They were only interested in the present state and what they could record at that time. They worked with fixed species and certainly not with larger groups, which were left for the systematists. This caused two approaches to studying evolution: genetics helping to understand micro-evolution within cells, and ecology looking at the larger group as macro-evolution. Only very recently have these two parts of the system been considered together, and it is too early to see clear outcomes.

The two approaches are roughly what Aristotle and Linnaeus called comprehension and extension, leading to distinguishing whole branches on the evolutionary tree, well within the spirits of George Gaylord Simpson and J. B. S. Haldane. These two were among the giant intellects of biology on each side of the Atlantic from the 1930s to the 1950s, looking for major trends in evolution. They studied the morphological and genetic aspects of evolution as an integrated system and were less concerned with the classification of organisms. Other greats such as Sewall Wright, R. A. Fisher, Ernst Mayr, Julian Huxley, and S. S. Chetverikov in Russia were also involved, desperately trying to quantify biological systems. Their data and their powers to compute were pathetically small.

When these biologists were students, at the beginning of the twentieth century, there were two concepts of the species. First there was Linnaeus' fundamental unit, based on Aristotle's basic unit of matter, the unchangeable gifts from the Creation. To put species into some applied context he grouped them into genera, but did not imply any relationships between the species of the same genus. The second concept did take this relationship into account as a main influence. Darwin saw it as vital to accept that closely similar species were related and had diversified from an earlier species. He offered the 'principle of great [morphological] benefit being derived from divergence of character, combined with the principles of natural selection and of extinction' (*The Origin of Species*, chapter 4). Now we can see this concept of divergence as an avalanche on the sand pile, a stimulus for changes in the way that ecology and genetics interact.

Throughout the twentieth century there were ever more sophisticated definitions of species. Sexual compatibility was still a favourite test. Sharing the same sets of genes became an interesting theoretical proposition, if difficult to prove. Eventually, the huge databases of genome projects have enabled reductionists to use them to make accurate distinctions between species. It follows from this that the definition of each species as a sequence database is a goal for the new century. The individual has a unique sequence, while for the species it is a cluster of them. I think there will still be a problem about how you define a cluster.

A favoured method of dealing with such a definition is called cladistics. It is a way of bringing together data from the likes of a genome or an anatomy manual of different species to plot the relationships between them. The plots are called cladograms, and as usual in science, other things being equal, the simplest is taken as the favoured model to test how different species or individuals are related. You can see how the relatives cluster together. A clade is a branch of a tree of life containing organisms descended from a particular common ancestor.

If there is so much uncertainty and ignorance about defining living species you can easily believe that there is more for extinct ones. To make it even more difficult, fossils don't engage in the sexual act, so you can't use that to test if they are of the same species. Also, most extinct species are not preserved. The 1992 Rio Biodiversity Convention agreed on the urgency of making an inventory of all the world's known species. How else can we start to monitor the loss of species and their destruction in dwindling natural environments? As well as giving strategies for the care and protection of life and the environment, the Convention set the task of assessing the world's species. Two years later a UN Biodiversity agency estimated that the total number of species is between 6,000,000 and 200,000,000. In reality it's hard to know how many there are, and some have been described and named more than once. Nevertheless 55,000,000 seems to be the most commonly accepted estimate.

Ignoring these difficulties, Dilshat Hewzulla and I have tried to calculate how many species have existed since the beginning of life. We had to make several really crude assumptions: that each new species lasts for 10 million years before it becomes extinct, that life began 3,500 million years ago, that there are now 55 million species, some known, most unknown. Perhaps it's not so outrageous, because there is plenty of evidence from different sources to support each of these statements. With these facts we calculated the number of origins and extinctions and ended up with the grand total of 1,107,807,741 species. And that's without knowing what the species are and without knowing where they were.

This crude exercise points out several of the difficulties facing different specialists in the new culture of biodiversity, bringing together data from molecular and morphological systematists, paleontologists, and ecologists. They have different understandings of basic concepts like 'species', 'evolutionary tree' and 'migration', with no sign of consensus. In each parent discipline there is a long history of argument about how the range of diversity is named and categorised, how they relate through geological time and how they moved between changing environments.

Some biologists believe that these fundamental concepts are getting us bogged down in semantics, leading to no useful ideas or practical conclusions. To help understand an integrated evolutionary biology these classical concepts might be replaced with very different ways of looking at things. Do different branches of evolutionary trees show distinct patterns of change? Is there a clear response from related species to the physical changes in environment? As soon as we ask such questions we are confronted with the difficulties of what the words actually mean. The more you investigate the meaning of concepts like 'species', 'groups', 'branches of evolutionary trees', the more you realise that different people understand them to mean very different things.

So it was with some interest in these matters that more than a hundred of us argued one March evening in 1997. It was at the Linnean Society of London, which first heard Darwin's essay about natural selection, read publicly in 1858. This time Darwin's portrait was staring down on the assembly and the bench seats were arranged to form a debating chamber, as in many learned societies during the nineteenth century when science was starting to be confrontational and conversational. Nowadays debates like this are very unusual.

The motion was 'This house believes that Linnean classification without paraphyletic taxa is nonsensical.' In other words, classification trees are different from evolutionary trees, the difference between Linnaeus' and Darwin's kinds of species. How we organise the names of organisms (taxonomy) can't reflect even their possible evolutionary pathway (cladistics); taxonomy is different from cladistics, taxonomists are different from cladists. The motion was carried by 69 votes to 43. Linnean classification lists won the day. In 1997, why should there have been such an interest in such a dull-sounding problem?

The argument won't go away, and the next salvo was won by the cladists. It was summarised in the 23 March 2001 issue of *Science* magazine, in an article entitled 'Linnaeus's Last Stand?' A fight has erupted over the best way to name and classify organisms in light of current understanding of evolution and biodiversity. More than in London, American cladists are in the ascendancy. Their 'Phylocode' system of nomenclature attempts to set a definition and constituency to each branch in the Tree of Life. This is partly due to the success of a website (*phylogeny.Arizona.edu/tree/life/html*) run by David and Wayne Maddison at the University of Arizona, which is attracting authoritative interpretations of many clades and aims to set some kind of 'standard'. A natural bedfellow for such a database is the DNA sequence information for the same branches of the tree. Supporters of Phylocode (*www.ohio.edu/phylocode*) have T-shirts boasting 'phyla.schmyla —

support rank-free classification'. The ghosts at Europe's natural history museums are rattling their chains.

At the Linnean Society debate, the traditionalists won the argument for retaining the classification of organisms with species, genera, Family, etc. In America, now more influenced by the internet, cladistics and DNA sequencing, I guess the result would have been very different. To polarise the issues like this however is very over-simplistic. The reality is that the system is very complex indeed, and tempting as it is to summarise our understanding of it, that does lead to confusion. I think that more than ever, we need a more subjective attitude to what is, after all, a very blurred outline of nature. Physics and chemistry attract quantitative laws and interpretations more easily than the messier world of biology, with all its associated disciplines from the natural sciences.

Old wine in new bottles

I began to question the meaning of the species concept as applied to the fossil record more than thirty years ago, through the stimulus of a long series of round-table seminars organised by my mentor Bill Chaloner at University College London. The table seemed to be more than 3 metres in diameter and the shape meant that we were all seated equally. Inexperienced voices like mine could join in with the others without the usual threats we expected in the debating chambers of those days. This democratic exchange of views stimulated wonderful argument and amusing thoughts and guesses. Bill and most of the others were quite happy to continue using species and genera for their fossils. 'The names are familiar, we all know what they mean, and if there is some uncertainty about the detail or the interrelations, then that can be put right. The differences are what scientists argue about, and we are pleased to disagree, like here around this table.'

But there was one man sitting at the round table who disagreed with this attitude to naming fossils. He feared the vagueness of the names being used, arguing that they lead to crude comparisons and imprecise age determination. This was Norman Hughes, who died in 1992. Although he was well known internationally and had a permanent fellowship at Cambridge, his ideas of challenging the use of species and genus names were rejected all round.

Norman was at the first international scientific meeting I attended. It was the 23rd International Geological Congress in August 1968, in Prague, during the Soviet Empire's invasion of Dubcek's Czechoslovakia. Troops and tanks had greeted us early in the morning, and the lecture sessions continued despite the delay. It was pretty scary, not knowing what was going on, not knowing how it was going to end, with columns of tanks and hostile troops parked in the streets outside as low-flying Soviet fighter aircraft screamed overhead.

The debate was about evolution, and how we monitor the changes by naming intermediate organisms on the branch of an evolutionary tree. Some of the great thinkers of paleontology were there, Jim Schopf and the host, Professor Frantisek Nemejc, both believing that living species and genera should be found as fossils back in the geological past, well before the ten-million-year duration of most species.

Jim was an archetypal Midwest American from the 1930s, determined to push his view. Professor Nemejc was the opposite, a quiet, frail man wanting to share the beauty of fossils and their meaning to life. Both men argued to retain the hierarchy of modern species and genera back in time as far as possible. This was to have the advantage of keeping some fix on changes from the present to the distant past. It was simple application of Charles Lyell's principle of uniformitarianism, 'The present is the key to the past', a principle that still rules most geologists' minds. The further back in time you take the modern groups the more power to the principle. Schopf and Nemejc were proud to have been brought up in this tradition.

Norman Hughes was having none of this and he said so. He spoke with a lucidity that stunned the still-frightened audience with about the same force as the Soviet troops outside the building. In fact, his argument had a lot in common with Dubcek's stand against the Soviet Empire earlier in the Prague Spring. Should we all profess loyalty to the official principle, whether it be Lenin's political ideology of communism or Lyell's principle of uniformitarianism? Hughes was far too English to conform to any of that, and after all, he was a fellow at Queens' College Cambridge. He had the confidence to challenge authority, the mischief to argue in the top places, and the intelligence to know when someone was wrong.

At Queens' Hughes was wine steward, surely one of the most wonderful duties imaginable. As well as tasting and learning about vinology there was the duty to spend thousands of pounds a year maintaining the life of the cellars. It also provided him with a brilliant metaphor to describe his attitude to naming fossils. Wine and organisms both have very complex composition and change through time. It's convenient to put both into containers and give them names and dates. Sometimes experts can recognise the characters of each, such as *Château Pétrus Pomerol 1945*, but often they can't, and processes like mixing are usually difficult to understand.

He argued that in fossil groups evolutionary changes mean that the different kinds of organisms keep changing and so they have a short duration as the same form. This means that the further back in time you go, the less likely you are to find modern species and genera. Hughes argued that to monitor evolution through geological time you need more than just a species name and a genus; you need the place and the time and the full description and good comparisons with other fossils of the same kind. This perfectionist technique is very slow and hard for others to follow; one simple name is easy to use even if it may mean different things to different people.

On reflection, my overview of Hughes' fussy stand is that despite our attempts to classify wines and organisms, no system can work reliably because each is too complicated to be described simply. Most bottles of wine that reach the shops, if not Queens' College, are non-vintage and would be hard to identify without the label. Certainly we can't use the same labels for different vintages, albeit from the same vineyard. The argument came up in many of his lectures.

As though from the back of the room, the roar of a Soviet fighter aircraft just overhead drowned the discussions, many of us fell to the floor in fear, and the blackboard vibrated from the engines' departure. Professor Nemejc composed us: 'Please don't let them disturb us. They don't know what they're doing, let alone why they do it.' That same dilemma faces a lot of biologists still. The Cold War polarised political attitudes and I think a comparable gulf is being forced by the role of environmental change with evolution upon biologists.

There is a compromise solution which was favoured at the Prague meeting by Peter Sylvester-Bradley, then Professor of Geology at the University of Leicester. He was well known to be concerned about how fossil species may be different from modern ones and gave some of his allegiance to the 'splitters' like Hughes, who were keen to give different names for lots of little groups within some sets of fossil organisms commonly identified as single species, while also giving support to 'lumpers', who push everything together when unable to show any clear and meaningful differences. Through the remaining years of the Cold War the dilemma of naming fossil species has been settled by adhering to one or other of these two extremes.

Through these last thirty years many paleontologists have seen themselves at one or other of Sylvester-Bradley's extremes – hard to be midway. But recently there is a shift from this search for clear ideas about species. The central stage of today's evolutionary theatre is occupied by entities instead of wholes, and by the conflicts between these two levels of complexity.

Methods of visualising evolutionary change

In the early 1970s different ways of thinking about how relationships branch between species were reflected by a moment of unease at the Natural History Museum. The museum is the shrine to Darwin's memory, the headquarters of work on evolutionary biology. The peace was shattered with the appointment of a new Director, Neil Chalmers, an educationalist from the Open University. He was quite a different kind of person for the job than was normal, or indeed, expected;

change was in the air. Working methods would have to alter, with more emphasis on presentation, on molecular systematics, on interpreting the enormous amount of material, and of course on raising money.

Meanwhile, the research council giving government money for related research received a warning that UK research in evolutionary biology was too descriptive, old-fashioned and unproductive. It seemed like a second threat to the day-by-day operations of research in natural history. The research council invited a group of senior paleontologists from the United States to meet some of their staff and discuss the state of the field with other UK specialists. It was held at an airport and became known as the Heathrow Meeting. They agreed that a lot of work being done was purely descriptive. One verdict was: 'stamp collecting will have to stop'. This was not real science because the normal methods of working didn't allow particular theories or interpretations to be tested. Instead, they relied on the whim or judgement of each particular observer. Implicit in the insult was that much more analytical attention was needed instead.

The timing of this criticism couldn't have been better for some of the staff at the museum: those determined to introduce cladistics into paleontology. The analyses can be used for objective scientific testing of relationships between morphologies, species and locations. There need be no more stamp collecting – instead this is real science. With cladograms in the exhibits at the museum, the spirits of the English biological establishment were shocked. The elderly specialists were already on the defensive. In addition to the threat from cladistics, another stranger loomed on the horizon: genetics.

The view that taxonomy and paleontology are not scientific, not testable, had become widespread from the 1960s. This coincided with the rise of molecular biology, which made earlier traditions of looking at animals and plants as whole structures less fashionable. Looking at dead specimens can't be science, but calling it 'stamp collecting' was a big insult to the traditionalists. Chalmers at the Natural History Museum knew he had to do something about it. That's why he was appointed – to cut out the philatelists and replace them with people wanting to be more experimental within evolutionary biology. So molecular biologists were brought in to start work on the big genome projects that are just beginning to produce huge sets of nucleotide base sequences from DNA in thousands of different species.

A sure sign that this new form of paleontology had come of age was revealed in *Nature* in 1984. Impressed with how evolutionary theories could be tested by cladistics, the UK's leading evolutionary biologist, John Maynard Smith, welcomed paleontologists to the 'High Table' of evolutionary discourse. We can use cladograms to connect different kinds of organisms together according to what they have in common. The features could be whatever you chose, usually some structural feature or pattern of behaviour. More recently DNA sequences produced a large number of similar graphical interpretations, while other cladograms could be created from data of different species' geographical distributions. The inter-relationships depend on which characters you choose, so you could test your theory by using different variables. Furthermore, these clusters could be compared statistically to give a priority to which set has most validity.

The cladistic method made a big impact on the way we interpret data both from conventional structural paleontology and from sequencing databases. But I fear that some of the early advocates prophesied it as the single solution to understanding relationships within evolving systems. In his book *Deep Time* Henry Gee has charted the events and developments in these early years of cladistics, recording battles between the establishment and the 'gang of four' from the British Museum, intellectual mavericks sharing their joys and fears in the pub over lunch. Twenty years on, the implications of cladistics have settled within the general practice of paleontologists. Cladistic techniques are now used alongside other ways of examining evolutionary problems, objective and subjective, experimental and authoritative. The cladistic revolution has been comfortably included into the whole investigative process of evolutionary biology.

But still, I think there's something wrong with the current ways of trying to understand evolutionary changes. There is so much emphasis on concepts such as competition, survival of the fittest and evolutionary progress that we may be missing something. Is the confusion between the classification hierarchy, the cladistic hierarchy and the phylogenetic hierarchy simply clouding our view of what evolution is really about?

Most people have a particular explanation for the mechanism to drive evolution. The first is creationism; the second, natural selection. The genes and DNA are just the carburettor and the sparking plug for this driver. Could other mechanisms exist? Can the system be in control of itself and its inner complexity sufficient to provide the driving force? Can we therefore ignore the role of the environment?

This is a further way of trying to understand evolution by getting away from the branching hierarchy altogether. It is to stand further back out of the shadow of this tree of life. Think less of species and genera and their inter-relationships, the small twigs on the tree, and look instead for trends and patterns of entire branches or clades. Look more closely at the whole rather than its parts.

Wolfgang Wieser, a zoologist in Austria, has recently suggested a radical model for thinking about evolution. He sees notions of cooperation, with some coercion, between autonomous parts as leading to more complex systems of biodiversity. Instead of the prevailing species-based approach to biodiversity, we can study the dynamics of communities based on how all the individuals get on together. Wieser's organisational levels consider integrating cell biology, the individual and the community, without being too concerned about the details of the particular evolutionary pathways. His model leaves most questions concerning systematic biology unanswered, but it does offer a different way of thinking about the very complex systems involved.

It's not a huge mental leap from Wieser's tubular model to Agassiz's spindles. They are both graphical metaphors setting out to help under-stand how sets of organisms evolve through long periods of geological time. Although they involve different variables, communities and members of communities and the same taxonomic group respectively, they display similar changes. Each tube and spindle appears to show the same basic shape. Once the new association has become established and the peak diversity reached, there is a long, slow, often impercep-tible fall in the size of the system. It is a shape that seems to crop up time and again in biology and ecology, the same bell curve that has been used to exemplify market growth and much more.

Early diversification was slow, giving time for evolutionary processes inside the cells to respond to newly developing environments. There was then a flurry of genetic activity causing gene sequences to recom-bine, mutate, and produce new proteins. Some of these were useless and caused failure, others worked well, especially in new environmental conditions. For example, just one small change in ocean circulation caused by drifting continents can change the terrestrial environment radically. It may be a small change in rainfall or some other aspect of the weather, a change in soil constituency due to a new rock erosion regime, even a new addition to the flora or fauna. In turn, these slight environmental adjustments can influence the biochemical activity of a particular enzyme. They can trigger a long sequence of reactions inside the cells, eventually changing the way DNA is used.

We know very little detail of such interactions between environmen-tal biology and cell biology and how they bear on evolution. Here is one possible scenario. When a new group originates, with a small number of individuals successfully invading newly available territory, and their new genome provides the biochemistry that best fits the new surroundings, diversification gathers pace. After a slow start, rapid diversification reaches a clear peak, followed by a slow, long fall in the range of diversity, leading to extinction. It follows that for large clades with greater diversity at the time of maximum expansion, it will take longer for that clade to become extinct. Nevertheless, it is inevitable that extinction will occur.

For the time being, if one resists the temptation to join the spindles together, or say that one is more closely related to another, the model has no need for hierarchies, or detailed taxonomy, or features moving from primitive to advanced states. Although the shapes of Agassiz's spindles have been drawn without precision, many diversify quickly once they are established. This represents the same processes of cell biology that occur in Wieser's model, and equally, when the ecology of Wieser's 'group' and 'society' has become established, the diversification curve is at its maximum. The final loss to extinction takes the longest time, hanging on with fewer species and individuals in smaller and smaller areas, which is why islands so often harbour the last passengers, at the end of the line.

Another new model for evolution

I showed Dilshat the diagram from Agassiz's 1833 work and illustrations from some of the many textbooks following the same idea of estimating the change in diversity of all the large groups of plants. Dill and I talked of the genetic processes around the time of origination, the delay before the environment had altered and enabled the genetic changes that had happened unseen at a cell level to show up in the organisms' structure. The sudden radiation led to an early maximum number of species. We had seen it all before through different parts of geological time. Dinosaurs peaked in the Late Cretaceous, mammals in the Miocene, more species throughout biology during the maximum warmth of the Oligocene. Then there was the very slow decline in diversity to eventual extinction, usually in some relict isolated location.

The new interpretation challenged Dill to write new computer programs to format the data and analyse it statistically. He devised a mathematical formula, a model, to test some of our databases. If the changes predicted by the model followed the trends of the actual data, then we could be sure that the model was correct. It meant that we would be able to use it to fill in gaps in the fossil record, and even to make predictions into the future. More important, we can use the model to help confirm which Families are part of the same evolutionary branch, and even part of any large group of animals and plants.

A working model might also help clear up the confused situation of unclear jargon (species, genes, etc.), opposing theories of evolutionary change (linear, punctuated equilibrium or exponential), opposing methodologies (hierarchical evolutionary trees, cladistics, etc.), and threats of extinctions created by external forces. With the raw material of our search in such a state, it is no surprise that we have trouble understanding it. We don't even know whether we can use computer analysis of large data sets to answer these questions. It might be a kind of virtual quiz show without anyone being too sure of the correct answers.

The model we worked out describes a very simple process to account for evolutionary changes within groups of animals and plants. It is defined by a simple mathematical equation:

$$\underset{\sim}{N}(t) = \frac{N_f N_0 e^{-\gamma t}}{N_0 + (N_f - N_0) e^{\alpha N_f t}}$$

which we have proposed to describe the spindle shapes in Agassiz's drawing. The equation integrates overall diversity (that is the number of Families present in the group at every one-million-year interval), the geological age, the number of Family records at the time of peak diversity, and a unique extinction factor which we had calculated from Agassiz's spindle shape. All these numbers can be downloaded from the *Fossil Record 2* database over the internet and fed into a Microsoft Excel spreadsheet file. You can do this automatically at *www.biodiversity.org.uk* by clicking on 'search data' and selecting *Fossil Record 2*.

We set about testing our model with information from each phylum in *The Fossil Record 2* database. When I first saw these curves I was bowled over by the same shape appearing in them all. Our first tests also showed the exciting result that the actual data from *The Fossil Record 2* fall right on the curves calculated by our equation. Then I remembered the simple changes showing up in the sand-pile experiment. Are our diversification curves following similar trends? Per Bak and his Los Alamos colleagues had lots of data from experiments with sand piles and cars on motorways to test their early theory. Now there's enough data from evolutionary biology to explore the organic world. Is each large group a complex self-organised system? In chapter 3 I showed how the entire database conforms to the patterns of self-organisation. Maybe the contained, individual groups of types do as well.

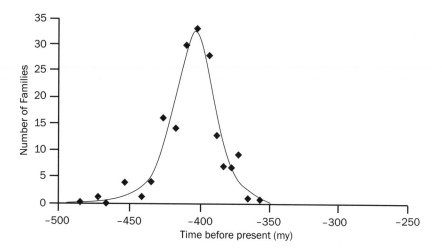

Fig. 5.2 The number of Families of fossil Agnathans occurring every one million years in *The Fossil Record 2* database. These Agnathans are jawless vertebrates like the modern lampreys and hagfish. The curve starts as an exponential and then reaches a peak according to the model in the equation on page 149. (original compilation)

The most obvious test of the spindle model is on the theoretical group of organisms that we know to be completely extinct. If the model is correct their diversification curve through geological time should give a complete spindle shape. We began with two clades: the jawless vertebrates called Agnathans, and a small group of funny little doglike animals called the Cimolesta.

The Agnatha first appeared about 370 million years ago and mostly became extinct 350 million years ago. Their Family diversity curve is shown in figure 5.2 with a symmetrical spindle shape leading to a quick decline before eventual extinction. The data are from very old sediments, from times when the fossil record is very incomplete, a feature often used to cloud over interpretations of data like these. Perhaps that's why the fall in diversity to extinction is quicker than the model expected. The further you go back in time the more cloudy

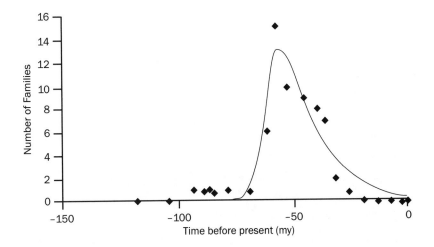

Fig. 5.3 The number of Families of Cimolesta occurring every one million years in *The Fossil Record 2*. The Cimolesta were small dog-like animals. As in Figure 5.2 the curve starts as an exponential and then reaches a peak, this time leading to a slower extinction. (original compilation)

the images are likely to be, yet otherwise our curve does follow the model very well.

Then there was the data from the other clade, the Cimolesta, from very much younger sediments and with a good record of fossils. These Families began to appear during the Cretaceous, peaked in the Early Tertiary, and became extinct just 3 million years ago. They were stimulated to diversify by the absence of dinosaurs. This time the diversification curve (figure 5.3) gives a perfect fit to the shape that comes from our model's equation. It fits our theoretical assumptions in every way, with delayed origins, a very fast diversification to the peak, and a slow reduction of diversity to extinction.

Most large groups of animals and plants are not yet extinct and will only give a shape to the basal part of the spindle. Once again, all those we have tested show a pattern of change that fits our model. Angiosperms (figure 5.4) and mammals (figure 5.5) are the largest groups

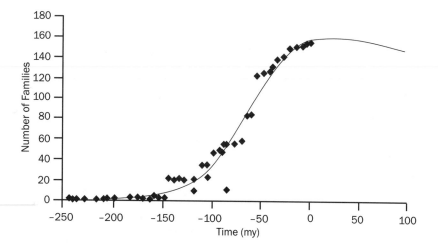

Fig. 5.4 The number of Families of angiosperms occurring every one million years in *The Fossil Record 2* database. The model equation on page 149 gives the solid line which predicts maximum Family diversity over the next 20 million years and then a very slow fall to extinction. (after Boulter & Hewzulla, 1999)

we have plotted, but others follow the same trend, as you can test yourself through the internet. Both groups showed their maximum rate of diversification through the Paleocene and Eocene, and the number of mammal Families peaked in the Early Miocene. The angiosperm Families are peaking about now. The computed curves can predict the modelled changes into the future, and if they continue to follow the model, mammals should be extinct in another 900 million years, and angiosperms much later. Both these estimates assume that the system is entirely self-organised without interference from outside.

But, of course, biology isn't a straightforward set of facts and principles and the curves don't always follow our simple equation. There are exceptions, and as usual they have proved to be even more interesting than the normal bell-shaped curve.

The first exception impinges on the topical debate about relationships

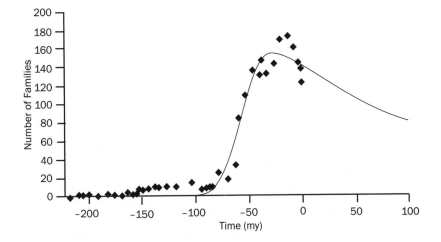

Fig. 5.5 The number of Families of mammals occurring every one million years in *The Fossil Record 2* database. The solid line is the prediction of extinction time from the model on page 91 if the Earth-life system is not influenced by external events. (after Boulter & Hewzulla, 1999)

between birds and dinosaurs. To understand the model's contribution to this debate, it's necessary to go back to the idea of exponential diversification I discussed in chapter 3. When we plotted the diversity of all fossil Families for figure 3.6 we found an exponential character to the curves. This is the same shape that our model gives as each separate clade, and every phylum, is now beginning to rise to reach its peak in diversity. The rise becomes very fast, as though moving to the exponential, and begins to fall when the time of ecological saturation is reached. The spindle curve is yet another expression of self-organisation within the clade, just as the exponential shape is manifest when the system is very large, as it is with all Families if they continue to follow the model (see figure 4.1).

Of course they won't follow the model. The model doesn't take into account the major episodes of evolution such as mass-extinction events. We have already seen the dramatic effect of the K–T event on the number of dinosaur Families (see figure 2.3). The two major dinosaur groups, Ornithischia and Saurischia, originate around the start of the Mesozoic. They both show three peaks at about the same times, and of course they both fall to zero at 65 million years.

When the bird Families are added to the dinosaur data there is a good fit to the exponential (see figure 4.1). But when just the Saurischia are added to the birds, the exponential fit is almost perfect. This must mean that the birds were part of the Saurischia group because the two build up a perfect exponential curve.

The second exception to the simple bell curve came first from our analysis of the diversity data for two different groups, amphibians and ferns. For both, our curves (see figure 5.6) do not have simple bell shapes but instead show two high diversity peaks, one in the Paleozoic and the second in the Late Tertiary. At first this seemed to be a major setback to our model, but then I started to think about the biological justification for these large groups over such a considerable length of time. Could the concept of each of these groups be artificial? So I split the constituent Families of each group into two, one with Paleozoic Families and the other with younger ones. We ran the program again, and this time the familiar bell shapes returned (figure 5.6). This exercise showed us that if we follow the spindle model for the evolution of species, what we call amphibians and ferns went through the full experience of evolutionary diversification twice. That is to say it must be for specialists with evidence from other sources to confirm this new idea.

The third way in which the bell curve and exponential models are challenged is with the intervention of external forces, mass-extinction events that disturb the normal self-organised system. The dinosaur Family extinctions at the K–T boundary (see figure 2.3) are clear examples, and the reduction of amphibian Families at the Permian–Triassic mass-extinction event (see figure 5.6) disturb the smooth bell curve associated with changes inside the living system.

So what do all these spindle-shaped curves mean? Could it be that the model's equation represents a real biological entity, a natural club

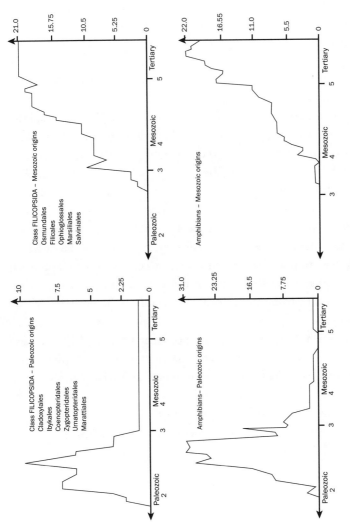

Fig. 5.6 The number of Families of fern-like plants and amphibians occurring every one million years in *The Fossil Record 2* database. The credibility of the model from the equation on page 149 is maintained if these two groups are split into two, one originating in the Paleozoic, before mass-extinction event 2, and the other just before or after mass-extinction event 3. (original compilations)

of members with the same origins? They share common genetics, geography, ecology and morphology leading to an evolutionary lineage of the same shape. The crucial feature of the model is its use of the idea of exponential diversification, a feature we first recognised by thinking of evolutionary change as a self-organised system. It is like the sand grains causing the pile to grow, leading to the inevitable avalanche. The model is also relevant to separate parts within the organised system.

These preliminary investigations mean that, although it can doubtless be made more accurate, the equation does describe the evolutionary changes that have taken place in these groups. What's more, the same equation holds good for most phyla, as we know them in the *Fossil Record 2* database. We have defined evolution within each phylum by one simple equation. For some time mathematicians like J. B. S. Haldane and Ian Stewart have suspected that biological shapes and processes obey mathematical rules. All regular shapes can be described by a single equation and the spindle-shaped curve is no exception. Its calculation matches the simplicity of the patterns that emerge. There is a natural beauty to be found in these wonderful summaries of evolutionary change.

Now, however, through a 500-million-year history of life on land, a unique change is taking place. For the first time a mammal species has evolved that can change the environment, and what's more we are changing it dramatically, quickly and selfishly. When habitats are lost species become extinct, and when that happens on a large scale through enough time, the avalanche effect becomes very powerful.

6

A Man-Made Extinction Event

Modern humans' first aggression

Some scientists believe that a swift and violent act of aggression brought modern humans into full control. About 42,000 years ago we migrated north from our centre of origin in east Africa, up through the Middle East, either turning right into Asia or left into Europe. There we soon came across groups of other humans who had arrived about 120,000 years ago. These other groups were the Neanderthals, and they kept their distance cautiously. The different groups didn't like one another.

It looks as if from the very first conflict we were the most aggressive. Our close relatives were much quieter and more peaceful beings, even though they looked very much like us and had most of our characteristics. We all had dextrous hands to make spears for hunting, and bipedal gait for moving in groups through the grassy bush of eastern Africa and the northern forests. But modern humans had other differences. Arguably, we were more intelligent, had better language abilities and were much more selfish. It follows that we are more likely to have been much better hunters, ate more and grew stronger.

It's very hard to piece together a few broken bones from a fossilised group of differently aged primates scattered over a desert or cave floor, and to be sure that they come from the same animal needs very special circumstances. It follows that the reliability of any description that attempts to recognise an actual species cannot be totally objective. Nevertheless, fastidious care by anatomists has defined more than 5 species of *Homo* that lived in east Africa over the last 2 million years.

Some are based only on fragments of skulls, others on very few specimens; usefully naming distinct needles in a haystack sounds like an easier art. The latest distinctions between *Homo sapiens* and *Homo neanderthalensis* are based on features of the early development of the skull. That's a classic criterion for distinguishing one species from another that leads us to ask once again what species really are.

The picture has been made to look even more abstract with recent detail from results of the first DNA analysis of humans. Some of these untried data point to a much earlier date for the origin of *Homo sapiens*, up to 465,000 years ago, compared to 130,000 years for the oldest known fossil. Other calculations from modern DNA make further controversial conclusions, declaring that there were always fewer than 10,000 individual modern humans in the population in Africa. If the figure is correct it starts the mystery of why there were so few of us many millennia after our origin. The same pressure may account for our surprisingly early migration north and how that effected a rise in numbers. For some reason we appear to have done less well in Africa than in Europe and Asia.

The most reliable parts of the fossil record are usually the youngest. They show that modern man was living in small separate tribe-like groups in north-east Africa no more than about 100,000 years ago and migrated north into the Middle East a few hundred years later. A westward passage opened about 40,000 years ago when most of Europe became thinly populated by modern humans within a few centuries. We shared the territory for the first few thousand of those years with the groups that had lived there long before, Neanderthal men.

These sad creatures looked as though they were used to hard physical labour. Their short bowing trunk and limbs had heavy muscles, a short neck and a large head. They reacted to the cold climate with a hairy body and a large nose to warm the air they breathed. (Others argue that the large nose is to dissipate excess body heat.) In comparison to modern humans, their small larynx suggests less developed speech and a large pelvis allowed larger babies.

Suddenly, all traces of the Neanderthals disappear from the fossil

gatherers theory, so for the time being these ideas are losing support.

In Europe and Asia there is much less evidence for large mammal diversity and extinction because most remains have been removed by the last glaciations scraping their presence from the surface. What little does survive is usually found in caves, where accurate dating is notoriously difficult. Nevertheless, about half the species are estimated to have been removed by modern humans between 30,000 and 15,000 years ago: mammoth, rhino, hyena, lion, panther and bears. Once again, the latest consensus explanation is that humans were the cause.

Man's aggression also explains what went on much further away, when man first colonised Australia and New Zealand. On all these islands, as well as Madagascar, the extinction of large mammals and flightless birds coincided closely with the arrival of modern humans. Although the hunters may have aimed for selected species, their destruction of the habitats meant that whole faunas of large mammals perished. But in Africa things were different, as we know from the unique survival there of big game. There's evidence for some large mammal extinctions between 10,000 and 120,000 years ago, but there are many fewer than on other continents. The reason is that in Africa humans and many of the large mammals evolved together side by side through over 5,000,000 years, so there was never an invasion by modern man. They originated in the same ecosystem.

The first human migrations out of Africa introduced man into new environments with new prey. This invasion very suddenly increased the rate of extinctions in Europe and Asia. Man's ability to fight in groups, our use of language and our resulting talent for networking all led to the fast success of our own species over others. These attributes mean that unlike other primates we have a unique relationship with the environment.

Although we have probably caused the majority of extinctions since the last ice age, there is evidence to support the view that changing climate has stimulated some plants and animals to migrate into unfavourable places, ending in their extinction. Examples are particular species of hedgehog, brown bear and beech trees that became isolated in southern Europe's few warm refuges during cold phases in the climate. Now these organisms have moved back north there are discernible genetic changes in the present-day populations that show up as new species in slightly changed ecosystems resulting from natural climate change. The scale of these changes is very difficult to measure. To distinguish the naturally induced from the human-influenced is even harder.

Migration is an important part of the evolutionary process for both animals and plants. It is a major cause of change, and introduction to new environments is the stimulus needed to start newly recombined genes working. Some good examples of this process have recently been recognised in human evolution. Early humans were dark-skinned, not black. That feature developed in Africans to protect them from the sun's radiation. Europeans' cereal-based diet led to a vitamin D deficiency and rickets, so they adapted white skin to allow the lesser temperate radiation to make more vitamin D. Siberian Mongols have small noses with narrow nostrils to stop them freezing up. Because these adaptations can happen easily and quickly, our species can migrate to new environments without becoming extinct. Other large mammals can make similar adaptations and survive. Nevertheless the finger of suspicion points more strongly than ever at man the hunter as responsible for many extinctions.

All life and all environments changing together are part of one huge self-organised system. As we are part of it ourselves, it is particularly difficult to recognise it and see how it works. The very recognition of such a single system does help our understanding of evolution. It means that we should adopt a philosophy that does not place humans at the peak of a pinnacle.

No response to climate change

Rarely a day goes by without a new climate-change story appearing in the media. These are not just about the weather and the impact of unusual or extreme episodes on our daily lives. The changes are put into the context of broad environmental issues such as the composition of the atmosphere, the height of sea level, the break-up of reliable seasonality and all manner of other torments. Politicians and lobby groups are taking the news stories very seriously, and the scientists who study these phenomena are winning large financial support for their work. International committees are giving reports of gloom and doom to their governments; tremors of crisis ripple down the corridors at the United Nations.

But none of this is new. All through the Tertiary there was frequent change in these parameters, and we know that they fluctuated on many different scales. In the context of geological history there is no reason to feel that the present levels of change are particularly unusual or, taken on their own, that they signal the approach of some major global tragedy. The causes of change that I discussed earlier, such as planetary interactions and the Earth's north–south balance, even sunspots, continue to govern the present. Likewise they have been responsible for the climatic events observed since the last ice age (see figure 4.3), and many of these extremes make our present weather look ordinary.

Before the Industrial Revolution extreme weather was not uncommon. There were severe winter storms and warm summers in the early 1700s, within the Little Ice Age between 1550 and 1850, when the Thames froze at London Bridge most winters. There were droughts and floods elsewhere further back in time, much in line with the fluctuating graphs of each environmental indicator that we try to measure since the last ice age. Changes in the weather and the climate are not unusual, and humans and other organisms have tended to take them in their stride.

Through most of the time since the last glaciation humans have been responding to climate changes in a straightforward and constructive manner. The first and most obvious response to the rising temperatures was migration – modern man into Asia and Europe, Paleo-Indians crossing the thawing Bering Sea land bridge. Our knowledge of the reactions to changes in climate becomes more sophisticated as the evidence gets younger. About 4,000 years ago vast swathes of the Middle East became very dry, forcing human migration. The alternative was to stay and encourage complicated irrigation schemes. Often these broke down and the people from different cultures forced together by the changes started to fight. Human aggression turned against our own species with the same ferocity that the Neanderthals had experienced earlier. The success of groups of humans coming together and cooperating to form tribes was good tactics for warfare, but bad for harmony within the species as a whole.

In South and Central America too, abandoned villages and agriculture are remains from similar human struggles against drought. Despite these kinds of setbacks, civilisation prospered, even gained from the disturbances between the newly emerging society and the changing environment. The more flexible the living systems within the changing environments the more the human populations grew. Human ingenuity was being put to the test and it thrived. It began to control the environment.

All the time, weather patterns were changing to affect human behaviour, although the rates of change did vary considerably. The archeological record is full of examples of such cyclical changes in intensity, location and detail. Civilisations and national character are partly influenced by local nature, and the interactions can account for trends in personality, language, immunity to local diseases and much more. Natural selection was working well within the complexity of different human ecosystems and cultures.

For other animals and plants there is no evidence of large-scale extinction over the last 5,000 years. Considering that rapid climate changes through interglacials like ours are such unusual processes through geological time, it's surprising that most other organisms have been able to survive. Yet this survival, mainly through cunning migratory strategies, is sometimes so false that it ends in extinction. One of the few species we know that behaved in this way is *Picea critchfieldii*, a subtropical Christmas-tree-like conifer. It couldn't adapt to a sudden temperature rise or migrate north to stay in the cold as its deciduous neighbours did. All known specimens are dead.

If our spindle-shaped model of extinction is correct it is going to be very difficult to be sure that a species is extinct, especially within a short length of time such as that between the whole glacial cycle, about 100,000 years (see figure 4.3). So it is with some scepticism that we consider actual numbers of extinctions published by the World Conservation Monitoring Centre in 1992. Over the last 400 years on the major continents about 100 species are reportedly gone, on islands 355. In order of highest loss they comprise molluscs, birds and mammals, insects, then a few reptiles and amphibians. Two things glare out from these results. The numbers of recently extinct species are very small indeed, and extinctions are far more likely to occur on islands than on continents. Whatever the scientific explanations, the figures for full extinction of species since the last ice age are very low. So what is all the fuss about in the media?

Extinctions in response to humans

At the beginning of the nineteenth century the world human population was around 900 million, and by the end it had doubled. Now, just 100 years later, it's getting on for 7,000 million. These two centuries began with the Industrial Revolution that led to unexpected pressures on the environment, the growth of cities and the indiscriminate burning of fossil fuels. Without thought or planning these human activities spread across Europe and North America to the new empires. Much the same imbalance of selfish human priorities continues today. Since the Second World War fifty years of economic growth in the developed world has considerably increased the rate of change.

Earlier I set out what seemed in prehistoric times to be the major distinguishing characteristics of modern humans: bipedal gait, dextrous thumbs, language, an ability to plan ahead, and a certain selfishness. How else could the European Neanderthals and so many large American mammals have become extinct so quickly? But to this list of dubious talents I must add one more skill: the ability to change the environment. For the last 200 years we have been doing it on a large scale, irreversibly. The changes we are forcing on one part of the complex system Earth after another are slowly leading to extinctions.

Extinctions are happening in three major ways. We are changing and erasing environments, we are introducing species that threaten previously stable ecosystems, and we are killing. It is clear from other self-controlled systems that once things start to set a new pattern of change in the timing of the system many old routines have to change, so that true recovery is not possible. That too is happening here and now.

Nowhere is the problem clearer than in the Caribbean Sea reef communities. Due to overfishing most large vertebrates had disappeared by the 1950s, but the reefs were unaffected because plankton-eating smaller fish and urchins survived, preventing the coral from being strangled by seaweed. In 1983 the urchins all died, and the algae have since taken over. They are now being joined by toxicity from fertiliser run-off, deposition of deforestation sediments and oil spills. Soon climate change will be adding its effect and finally end the life of the reef. Such reef systems are among the oldest ecosystems, going back more than 500 million years to the Early Paleozoic period. Older still are the blue-green algal boulder-like domes called stromatolites, best-known at Australia's Shark Bay. They are some of the oldest life forms and originated about 3,500 million years ago, before organisms with proper nuclei and membranes in their cells. All these ancient communities are exceptionally sensitive to change. Their very existence shows that their ecosystems have been stable for all this time.

One of the more politically charged examples of change in land use is our conversion of tropical rainforest to agricultural land. Similar human forces converted temperate forest in Europe and grassland in North America to agricultural land and the process is simply being repeated in the tropics. The effects are that few habitable parts of our planet have the same landscape now as before man wielded the knife. Because it has happened so quickly we have no full experience of how to look after what little natural landscape is left. Just 2 per cent of the original forest remains in north-east Brazil, and what is left is badly managed and heavily hunted.

Three-quarters of the tree species are dispersed by birds and mammals, usually by them carrying the seeds. Neither vector works effectively in false islands, small copses surrounded by open fields, where shade is much reduced and foraging and other normal habits are restricted. There is an intimate balance between shade tolerance, fruit size and wide or narrow bird gape of the beak. Each controls how much the birds eat. The slightest change upsets this balance. Rapid change in this scale means that the balance is not sustainable. The forces controlling the species–area relationship will eventually have their impact on these sad remnants of rich biodiversity. Many species of animals and plants are being brought to the edge of extinction. On the continents, large and medium-sized mammals are the most demanding and the most at risk, but the loss of so many plant species has an effect on their dependent insects and small animals.

Many more of the extinctions are on islands, where the last species of a group typically seeks refuge. The Galapagos, the Canaries, the islands of south-east Asia, have more than 10 per cent of all the species listed officially as being under threat. It seems however that worse still has gone before. Recent excavations on many Atlantic and Pacific islands show that most have lost half their bird species since colonisation by humans. Mammal species have been substantially reduced in number as well. Change of land use and hunting are the main reasons for the changes.

A recent example of island extinction, close to home, is that of the giant deer or Irish Elk. These handsome creatures had shoulders more than 2m high and an antler span of 3.6m; they populated Asia and Europe for 400,000 years. They didn't like the ice ages, less because of the cold than because the permafrost reduced the amount of food available, but they did survive. Soon after the last glaciation all individuals in Asia and mainland Europe vanished, forced westwards not by climate change but by human hunters. At last they found refuge on the island of Ireland. Recently, the youngest known elk remains have been found on the Isle of Man, just 9,000 years old, older than the earliest evidence of humans on either island. The bones show that the last individuals had smaller bodyweight than usual, suggesting that the small island wasn't able to provide enough food for these large beasts. Although their flight west was caused by man, the final nail in the coffin came from the ecosystem.

The first species known to have become extinct by man introducing an alien species was the dodo. This originated as a pigeon-like bird from the African mainland which had flown to Mauritius, a peaceful island with no predators. There was therefore no need to fly, so the bird evolved into a dumpy flightless thing about the size of a turkey. Dutch colonisers visited the island about 400 years ago. The visitors actually left the bird alone because it tasted so bad, but they brought their dogs and pigs and the ships had rats, so there was competition for the dwindling food supply. Dodo eggs tasted good to the new arrivals. The species was extinct by 1700.

Many other relict species of flightless birds found their way to islands, and most are now extinct for similar reasons to those that killed the dodos. The heaviest bird ever known, the 100-plus kg *Genyornis*, was also unable to fly and restricted to Australia. It was among over 50 species of vertebrates that became extinct just after man set foot on that island.

Another flightless bird, the elephant bird, 3m tall, sought refuge in nearby Madagascar until it became extinct about 200 years ago. Once again, there is debate about whether its demise was caused by human hunters or the limited food on the restricted space of an island. It seems plausible that *Genyornis* and the elephant bird were part of the same group of birds roaming the old southern continent of Gondwanaland. This is now split up into Australasia, South America, Africa and India, as well as the islands in between.

The Dutch settlers in Mauritius were among the first humans to introduce alien species to a fixed island community of animals and plants. Since then, we have introduced thousands of unfamiliar organisms to different parts of the world. The consequent pace of change is quite out of control, with the horticulture, agriculture and travel industries mixing things up indiscriminately. There are rabbits in Australia, Japanese knotweed in Europe, while floating pennywort from aquarium shops and garden centres blocks up the canals. GM crops are another form of human intervention which we don't yet understand, but then, since we invented agriculture thousands of years ago, we have created our own conditions to breed special crops different from the wild type many times before. Over a few hundred years we have changed the distribution of most of the planet's animal and plant species. Over a few thousand years we have completely changed the genetic character of most animal and plant crops.

The latest shock story of island extinction is closer in space and time. The island is Central London, the time is now and the species is the common house sparrow. There were thousands of these little birds in London's parks and gardens in the 1950s. Then the decline began and now there are few if any. An even more curious thing is that their population in Paris has stayed the same. We don't know why but I bet it's something to do with human beings.

In the UK, attempts to legislate against importers of aliens are unlikely to keep the invaders out of the wild, for once out how can they get back? Zoos and botanical gardens are usually responsible institutions, but they have released some aggressive creatures into places where they cause havoc. From America, the Red Swamp Crayfish threatens the British species; mink (imported by UK farmers) threaten the rare water vole; and the ruddy duck threatens native species. New Zealand flatworms have been eating up native British worms for the last thirty years. Of course, new examples come along all the time.

How can we begin to estimate the size and rate of species loss through these human intrusions into nature's way? The evolutionary model discussed in the last chapter would expect the final demise to take a very long time as declining numbers occupy smaller areas. Habitat loss causes greatest loss of individuals, but the final extinction of the few surviving individuals is protracted, perhaps for millions of years in just a few places. These relics are mentioned frequently in natural history books because they are so rare and sad. The Galapagos animals are best known, but ancient plant lineages have relics like redwood and tree ferns which have dwindled to tiny vestiges of their former widespread glory. It's on these observations that those designing the hotspot monitoring project mentioned earlier base their methods. By comparing the rate of species loss with the amount of habitat loss, alarm bells are designed to ring in the offices of environmental planners when things get too bad.

The same trend of long-drawn-out survival of the final relicts has been further considered by Bob May's group at Oxford, particularly Sean Nee. The Oxford group are vociferous wailers of gloom and doom: 'Extinction episodes, such as the anthropogenic one currently under way, result in a pruned tree of life.' But they go on to argue that the vast majority of groups survive this pruning, so that evolution goes on, albeit along a different path if the environment is changed. Indeed, the fossil record has taught us to expect a vigorous evolutionary response when the ecosystem changes significantly.

This kind of research is more evidence to support the idea that evolution thrives on culling. The planet did really well from the Big Five mass-extinction events. The victims' demise enabled new environments to develop and more diversification took place in other groups of animals and plants. Nature was the richer for it. In just this same way the planet can take advantage from the abuse we are giving it. The harder the abuse, the greater the change to the environment. But it also follows that it brings forward the extinctions of a whole selection of vulnerable organisms.

If humans were to fall into this vulnerable category, we too would become extinct. The effect of this would be that the abuse would stop being inflicted and peace and quiet would return. It would take several thousands of years for this to happen, and even longer for many different new ecosystems to reach a steady state of climax. Meanwhile, of course, evolution would set to work and increase the diversity of the newly selected forms, without the threat of humans and all the other species that our extinction event killed off.

To help us start to work out how to respond to this scenario we need to further question how we understand these complex problems. Do we give up attempts to slow down climate change? Do we just regard extinction as inevitable? If thousands of species of birds are made extinct by man's actions, so what? Other twigs on the tree of life will emerge, other species in the same clade will occupy the niches left vacant.

Regardless, scientific inquiry will go on. All attempts to gather evidence and propose solutions must be on a global scale. One such project aims to paint a global biodiversity scenario for the year 2100 and comprises scientists from eighteen different institutions around the world. What will the distribution of species look like in a hundred years? What are the most important changes taking place in global environments to influence the patterns? That group of scientists argues that change in land use will be the single biggest threat, followed by the climate change, the effects of nitrate fertilisers, the introduction of non-endemic species, and finally the increase in CO_2 in the atmosphere. They have argued that changing tropical rainforest to agricultural use and the removal of natural vegetation around the Mediterranean are among the greatest threats. The changes are imperceptible in the daily lives of local people: they are more subtle than by the knife.

Unproven fears

A big warning was given. In 1972 George Kukla from Columbia University, who studied glacial deposits in the Soviet Block, wrote a letter to the then President of the United States, Richard Nixon. It was to share his hunch from work in Czechoslovakia that the next ice age was coming much sooner than anyone had predicted and much more suddenly – perhaps within just a few decades. Nixon's science advisers dismissed the tip-off, and some have lived to regret that decision. At the time the entire subject was absent from the political agenda, with no public concern and little scientific effort monitoring the changes.

There is more and more new evidence from the Atlantic and Pacific Oceans that big climate changes are on their way sooner rather than later. This is because the temperature of air all over the northern Atlantic Ocean and north-west Europe is considerably higher than elsewhere at similar latitudes. The Gulf Stream plays a central role in maintaining this global imbalance. It has been like this since the last glaciation and it was like this during the last interglacial. The fear is that global warming will change this equilibrium and dramatically shift the circulation patterns, reducing their geographical extent.

This is what Kukla twigged in 1972, and now it's a very real danger. Some specialists say it will come in a few generations' time. But what is clear is that the global system of ocean circulation and its effect on climate change is very complex, and we are only just beginning to understand it. It appears that there are two large centres of power that to some extent control ocean water circulation. One is called the *North Atlantic Conveyor* and turns round the cold deep water and warmer surface water in the Labrador Sea. The other is in the southern hemisphere, in the Pacific Ocean, and is called *El Niño*, caused by another circulation system. There is evidence that sea levels rise in the Pacific, to the west of Ecuador, and that weather patterns are altered. Some years it doesn't happen.

El Niño is another large-scale oceanic weather system that we don't really understand. It may be a southern hemisphere contribution to the whole global system of climate regulation within the oceans or it may be a separate system influenced by atmospheric circulation. Both the north and south systems threaten icy temperatures, floods, and other bad weather. The latest fears are that the *Conveyor* will divert the Gulf Stream and convection currents to change the regular pathways of heat exchange in other oceanic waters. Instead of melting polar ice, which is happening now, that means temperatures will fall a long way very fast. It will be an event enhanced by global warming from human activities.

Present-day politicians are beginning to take these things more seriously than those of the Nixon era, as environmental problems become more important items on the agenda. But there are no votes to be won for the seriously unpopular political solutions that are required. Nevertheless, at the turn of the last century the prime ministers of Norway and Britain talked about the threat from the *North Atlantic*

Conveyor. They had been advised just how serious it is becoming and they decided to take some action. Civil servants were alerted, funding agencies told to set up special projects, meetings were called between specialists.

More than two hundred of us attended one of the meetings, at the Geological Society in Piccadilly, registering at a desk below an oil painting of the Piltdown Man discussion staged in the same debating chamber seventy years ago. The contrast with this earlier man-made confusion is paradoxical. Then, one man tried to mislead the world with one of the most convincing hoaxes based on artificial fossil evidence. Now, selfishness is causing us to ignore the scientific facts about climate change.

As is usual now at these public planning meetings, experts review the present state of knowledge, others suggest new methods and approaches, and after lunch everyone joins little discussion groups to share ideas and brain-storm new plans. If anyone had a really good idea, I doubt that they would share it; rather they would keep it back to use in their application for funding their own research proposal. Despite that, it's the best way we have to approach new research challenges like climate change, and a good way of spending a day.

The *Atlantic Conveyor* experts are from meteorology and oceanography and have spent their careers working to try and find patterns in the changes going on in the Atlantic Ocean. In comparison to the complexity of what goes on under the sea between Europe and America, what they have to work with and go on has been trivial. The ocean is influenced by ice, atmosphere, river run-off, wind and unknown sea-floor currents. It is impossible to make predictions without a lot more data to show how these things change through tens and hundreds of years. Only then will the experts be able to make better models to predict future changes. New research projects lasting five years will be planned from meetings like these to give the necessary data. Specialist observers of the ocean changes will collect new data enabling the computer modellers to try new estimates of what is likely to happen next.

The alarm about potential changes in the *Atlantic Conveyor* comes from recent analysis of convection in the Labrador and Greenland Seas, where the warmer surface waters from the south sink deep into the ocean, cool and turn southwards again. It seems possible that some time soon global warming will interfere with the formation of these water masses and suddenly change the way in which they affect global ocean water circulation, and thereby the pattern that controls our weather systems. The presence of more freshwater produced by melting ice has the potential to change the ocean density so that mixing shifts very suddenly somewhere else. The fear is that this may change the Earth's ocean circulation patterns and bring forward the time of the next glaciations by suddenly reducing temperatures. This was expected to happen about 40,000 years from now, but some computer models show it starting in only a few hundred years or less.

Once again, because we are living through a time of fast exchange of information from one established discipline to another, there are great breakthroughs in understanding. However, the news that emerges is not good. It is that most of the world's advanced human culture will be forced to migrate towards the Equator in a few generations. Worse still, much of the equatorial biodiversity there is becoming extinct through flood and farming. There will be nowhere for us to go.

The potential for a different but comparable environmental disaster lies in calculations from another much simpler system, Himalayan ice. Glaciers there are melting, and the freshwater run-off forms lakes that eventually burst open. These catastrophic floods used to happen once a century; now they take place every year or so, killing hundreds of people and causing great damage to communities and nature. One authoritative prediction is that at the present rate of climate change all Himalayan glaciers will be gone by the year 2035. Meanwhile, the billions of people in southern Asia who depend on the summer run-off and monsoons from the ice system will be faced with extreme shortages of water for drinking and for crops: a major disaster, waiting to happen, that we can't stop.

Then there's the Greenland ice that's melting. Some calculate that frozen on the surface of that island there is enough freshwater to raise the world's sea level by 6 metres. Predictions about the effect of the increasing flow of warm air on this are very difficult, but a dismal clue came from the Greenland Ice Core. It suggests that during the last interglacial, when temperatures in the far north were higher than now, the Greenland ice sheet was much smaller. We also know that sea levels then were 3–5 metres higher than today. The world's other large ground-based ice sheet is on Antarctica and is also melting faster now than since records began. This contribution of more freshwater to the world's oceans also has a noticeable effect on the balances of warm/cold and high-/low-density columns of water. In the last interglacial, they would have had a role in contributing to the increase, no doubt through the complexities of the *Atlantic Conveyor*. It seems reasonable to expect global sea levels to be on the increase in a big way, both from natural and from man-made causes.

Most of the world's high densities of human population occur near sea level, and most are beginning to build elaborate sea defences. Just a few hundred years ago cities like Hamburg, New York and London didn't have river embankments: they flooded as Bangladesh does today. In England it will soon be impractical to maintain the more remote defences and the sea will stretch down to Cambridge, as it did before man drained the Fens. The Netherlands will face a crisis, the great deltas of the world, at Cairo, Louisiana and elsewhere, will become submerged. Electricity generating stations, docklands, half the North's infrastructure will slowly be taken out of use.

Such tales about global climate change and their impact on our familiar landscapes are being told as a major political issue. Away from wars, it's hard to think of any other story that runs and runs. But this is a world war. Sometimes there is a clear human cause to the misery, while at others there's still support for the tradition that it's the way nature works. Most experts feel the changes are getting out of hand irrevocably, but it's very hard to be certain with clear proof. The sheer speed of the changes that built up last century makes monitoring easier, though clear and indubitable evidence remains elusive. The timescale for good evidence is just too short.

Since Nevil Shute's novel about a nuclear holocaust, *On the Beach*, written in the 1950s, there have been countless tales of the end of man on this planet. Now, it seems, the joke is that we are doing very well on our own, just with our use of fossil fuels. There is no need for nuclear weapons or the inventions of science fiction writers. It is our own aggressive selfishness that has led to our lifestyle, and this has evolved its own political system to maintain the status quo. Now it's too late to change and we cannot organise ourselves to stop. I speculate that our system is in free fall, out of control.

7

Humans and the Future

Another view of the future

Just across the road from the west door of Westminster Abbey is the Central Hall, an early twentieth-century cathedral built as the headquarters of Methodism. I think the architects felt competitive with those who built the ornate Roman Catholic cathedral in nearby Victoria Street, as well as with the ancient Anglican Abbey opposite. In better days for the chapels the preaching of Wesley's successors rang out within its cavernous mantles. More recently the orators of the day, politicians and philosophers as well as theologians, spoke in this forum for radicalism. What a contrast, when on a cold February evening in 1999 the same platform was taken by two proud atheists, Steven Pinker and Richard Dawkins, agreeing that science has killed the soul.

Pinker is an American psychologist much influenced by the developments in genetic coding now taking place in mainstream biology. His thesis is that the human brain is some kind of computer, shaped by natural selection and giving rise to mental activity. It means that a lot of human behaviour evolves much like the physical characters so familiar to evolutionary biologists. Pinker's world comprises the equivalent specialists, evolutionary psychologists, though they are still to prove themselves to most biologists. One of the first who has signed up to evolutionary psychology is Richard Dawkins, whose concept of the selfish gene caused such a stir in the 1970s. Dawkins believes that all that genes want is to be passed on to the next generation where they will have the best effect. This led him to suggest behavioural tasks for a comparable inheritable unit, the meme: a unit of cultural inheritance explicitly not biologically transmitted. But Pinker and Dawkins still lack evidence for their ideas, and this, together with their support for reducing explanations to simple yes–no alternatives, attracts the rancour of the social left.

So with his strong sympathy with Richard Dawkins' arguments Pinker looks for genetic adaptations of psychological characteristics. This kind of reasoning leads to the goal of explaining human behaviour by some process like genetics. It would be an objective mechanism, based on atomic units like DNA and the genetic code, and would be transmitted from one generation to another. Presumably it would evolve as part of the comparable biological process. The evolutionary psychologists are busy trying to find appropriate units of behaviour that they can monitor to help prove these new ideas, though so far no such evidence is forthcoming. But Pinker and Dawkins are not without their followers, as the audience at Central Hall clearly showed.

The basic idea that psychology is one step on from biology in the sequence of mathematics to physics to chemistry to biology came from Darwin himself. In *The Origin of Species* he sets out to confirm the optimistic supposition that psychology may have a 'new foundation', so extending his thoughts about the scope of natural selection. Are basic features of human behaviour, such as selfishness, characters that have evolved by natural selection? For Darwin it was clear: 'natural selection acts solely by and for the good of each [species].' The most prominent contemporary supporter of the myth of the meme is E. O. Wilson, the grand old man of American biology. His book *Consilience* even attempts to justify religion within the mantle of evolutionary psychology. Without evidence, the objective biologists have never been more imaginative.

Their argument goes something like this. Humans are self-conscious beings and so we can challenge our place in the environment in ways that no other part of nature does; we are creative and so this feature may be controlled by a code like the triplet one of molecular biology that selects amino acids in protein synthesis. This is what the sociobiologists Pinker and Dawkins are looking for so enthusiastically, to discover how animal habits are inherited. Their ideas are very attractive to the behaviourists because they believe that the mechanisms of human behaviour are contained within the organism. They think that this is a more powerful explanation than anything learnt about social and environmental influences from outside the cell.

That Dawkins and Pinker were asked to take opposite sides in a debate about sociobiology is a strange thing, because their books preach the same lesson. For them, the physical sciences work in the same way as the biological sciences, which include psychology. They are saying that an objective approach to this young subject shows it has come of age as a respectable science. If DNA sequences can contain all the information required to control biochemistry and behaviour, then all human life is programmed inside the cell. What's more, the programs can be replicated from one generation to another, and run functions as diverse as respiration and religion.

Their gospel is that the organism is the complete system, and they agree that evolution takes place within the organs and tissues, the cells, the nuclei, the chromosomes and the genes. What happens outside the organism seems to have little or no part to play in their arguments. By mutation and sexual recombination the system develops from within the organism and new forms of life evolve. The argument is stimulated in the new knowledge that the genes are at the centre of an organism's system, controlling internal structure and function. Since this standard system is universal in biology, why can't a compatible mechanism run the world of evolutionary psychology as well and then go on to explain patterns of behaviour?

This explanation of how life evolves represents one limit of a spectrum passing from high objectivity at one end to low subjectivity on the other. Genes are at the one end, self-organisation and chaos at the other. The first option is logical with lots of new molecular evidence, the second is based on hunches for explanation and the patterns of self-organised systems as evidence. Advocates of the second option say that evolutionary biology is such a complex system that simple rules like $E = mc^2$ can't explain it.

So a big question mark hangs over which approach to these things we must take, or how to make compromises in between. Do we follow Dawkins and Pinker in adopting a rigid quantitative approach and look for fixed logical answers to the problems, or do we admit that for the moment there are question marks hanging over the problems, however they may be defined? Should we put some of the mess surrounding the issues into temporary packages and never expect them to go away?

As long as evolutionary biology is dominated by highly objective perfectionists, most attention will be directed to the new methods of storing and communicating the huge amount of data for molecular biology. This is the new industry analysing the information in DNA sequence databases. Big money will continue to be spent on making more and more of this computerised information, in the hope that mankind will be saved with more accurate medical provision and by cheaper food.

Meanwhile, the darker sides of human behaviour that led to so many extinctions of other mammals just a few thousand years ago go unheeded. The evolutionary psychologists are well set along the objective approach and reject any suggestion that the whole Earth and life system may be in control, not just one part of it.

But that most threatening of all human characteristics, selfishness, rises time and again as the fundamental explanation of what we have been doing to the environment since the Industrial Revolution. The sociobiologists however talk a lot about an opposite, altruism, which some believe to be a feature that can be monitored to show evolutionary changes. They think that humans succeed because we help our fellow men. I do not share their optimism, for these wise forecasts about group behaviour ignore my sense of what the evidence is saying, that we are only really interested in ourselves and our close family. We will continue to burn natural gas to keep us warm, kerosene to fly us away for a holiday, and once there we will pump water to a swimming pool in the desert.

It was with these thoughts and uncertainties that I set off to Central

Hall Westminster to hear Dawkins and Pinker. I was not expecting to hear any answers but went interested to catch the ambience. Methodist Central Hall has 2,300 seats and I'd tried to book by phone a week earlier, only to be told to pick up a ticket at the door. As it turned out, they didn't have one – the place had been sold out weeks before. There were crowds of young and old, suits and anoraks milling around the entrances, and a queue of hopefuls waiting for returned tickets. The building can rarely have been witness to such a crowded debate about the soul for many a long year. There was a strong feeling of expectation, a buzz of camaraderie, warmth, a new kind of congregation. But what or whom were they worshipping?

Alone, I left into the cold night and a strong wind from the north, past Whitehall to Horse Guards Parade and The Mall. Everywhere was deserted, no people, a ghostly silence, my footsteps crunched on the frosty pavement. Not so long ago here, Diana's funeral procession had been the focus of the world. Now the square was empty. Without people the streets had a ghostly feel, as though they had no place in nature, just a forced environment. Through the crisp air it was becoming clear to me that our attitudes to the evolution of life on our planet are becoming polarised. These buildings, empty of people, offer the security of government over nature. But there and then the people were gone, leaving a sculptured scene of man's work. It had a distinctly artificial appearance, like a Dickensian image on a Christmas card.

Perhaps it was that cold realisation of the vulnerability of the human condition that made me see a significance in what Dawkins and Pinker were saying. If human behaviour really can evolve then there is surely hope for the future of man – a way through the impasse of environmental destruction that the world's politicians find hard to prevent. Such a cultured evolution of human behaviour would mean that something in the minds of government mandarins in these empty buildings could work out a global solution to the environmental crisis and the human condition. It would be something our present memes cannot do: pulling a new behaviour out of the bag, some as yet unknown way of enabling man to survive and prosper. These characteristics of the memes would change our patterns of human behaviour to suit their adaptation. Natural selection would favour these new patterns of super-human behaviour and we would survive.

But if human behaviour cannot evolve, the response to fast changes in the environment will be very different. There will be no reprieve, no stopping the progress of mass extinction, and man surely will be a victim within that. Our most damaging behaviour is selfishness and aggression, and unless they can change rapidly there is no hope for the ecological destruction to be halted. Our power to do damage has grown to make our aggression terminal, not just dangerous.

If the Earth-life system really is in control of itself, perhaps there is nothing that we or anyone else can do to slow our abuse of the environment. On the other hand, could it be that the system itself will see to it that the abuse stops? The damage we do to the environment causes many species to have difficulty fulfilling their own peculiar requirements for living. This means the resulting extinctions have hap-pened much faster than is predicted by our spindle-shaped model. Just as happened with the decline in dinosaur Families 65 million years ago, so now, Families of large mammals are becoming extinct at a very fast rate. This is instead of the slowly protracted fall in their diversity which was shown by our curve of changing mammal Families in figure 5.5.

The new idea: self-organised mass extinction from within

So modern man is kicking the sand pile and causing a severe avalanche that only started to crash down at the end of the last ice age. There is no telling when the slide might end because the fundamental cause continues: human aggression. The first phase was our killing other mammal species when we first encountered them, then through human history our killing of one another.

Since the industrial revolution there has been a distinct third phase, our continuing destruction of the natural environment. Most of what we hear about climate change and global warming can be ascribed to this human activity, though of course some of it is a consequence of natural cycles such as sunspots and other rhythmic processes outside the Earth. The consequences are becoming very apparent even as we register them on the scale of our single human life. If something like a new meme is going to change this human behaviour it had better hurry up and evolve quickly.

There is another argument that catastrophes like human-induced environmental change are a necessary feature of the self-controlled system of life on Earth. As with the avalanches in the sand pile, they happen when the system reaches a critical state and passes over the edge from one kind of world to another. Equally, extinctions are an essential stimulus to the evolutionary process. We know it from the patterns that show up in our curves of evolution from the fossil record. The exponential line of diversification must never reach a vertical.

That critical place between extinction and continuing life has a very simple manifestation within an organism's routine. A mouse might choose to turn right into the jaws of a waiting cat. Or it might choose to turn left. Famously, in June 1914, Archduke Franz Ferdinand's Sarajevo driver wrongly turned right into the path of an assassin. Some say that the history of the last century would have been different if the car had instead turned left. Each story is about an individual, but for that to become significant it must reflect the trend of the large majority. To move across that critical point, from one state to another, requires support from the mob of followers, whether they be mice, cars, sand grains or voters in democratic elections. Once on the other side of the critical point, in that other state, there is a very different atmosphere. In evolutionary biology, with a new environment, other species are waiting for their first re.l chance to show off the new features coded in untested new combinations of DNA.

The system of life on Earth behaves in a similar way for all its measurable variables, whether they are communities or ecosystems. For sand grains, substitute species or genes. For avalanches, substitute extinctions. Power laws tell us that large avalanches or large extinctions are much less common than small ones. The controlling factors for the sand piles are weight and angle of the sides of the pile; for mammals they are space and food within the ecosystem. We can kick the sand pile with our feet, and we can reduce the space and the food by changing the environment.

But what would happen to the life-Earth system without these external changes? Could it be like a pile without avalanches, eventually collapsing into a mess of white noise? The answer lies in our theory of exponential diversification within macro-evolution; the curve ever rising towards the vertical when the *Fossil Record 2* Family data are plotted (see figure 3.5). The situation starts to become critical when numbers rise above a comfortable quantity, whether the system is a pile of sand, cars on a motorway or large mammals in America. If there were no mass extinctions, that exponential curve really could have risen to the truly vertical. It could have happened long ago, and it could happen again if there were no extinctions holding it back from the vertical. If that were so, all life on planet Earth would cease. It would need to start again from scratch.

But that would be impossible. For example, when a teacher cleans the blackboard of the lesson's writing, specks of chalk dust are reflected in the rays of sunlight pouring through the windows. There's no way the dust can be put back into the writing on the board, let alone into the stick of chalk. Could this one-way process, entropy, be like evolution facing the exponential? There is no going backwards, only forwards. To stay still is impossible. As with the chalk dust and the universe leading towards higher entropy, it may be true to say that within the history of life there is forever the unrelenting trend to less order and at the same time towards greater complexity, until bits of the system reach a critical edge.

While life manages to develop complexity inside the triumph of entropy, this same growth of complexity makes the system less robust, more vulnerable to intervention – the chaos cliché of the butterfly's wings leading to a hurricane – than its technically much cruder predecessors. As we shall see, for humans at this critical point, there is no going back.

Towards human extinction?

The idea of imminent human extinction was a challenge to my research group, and we quickly set about testing it. The obvious place to start from was the *Fossil Record 2* database that had provided a basis for so many of our earlier ideas. As usual we fed the numbers of mammal Families at every million-year interval into a Microsoft Excel file to give a curve of diversification over the last 200 million years (see figure 5.5). We were becoming hardened to getting these summarising curves out of the database, and this time we were not surprised by the results. The exponential curve coming from the dinosaurs and birds, the first bell-shaped curves, described at the end of chapter 5, were genuine surprises, even shocks. But for decades we have known that the mammals' Family diversity peaked in the early Miocene, and our curve merely confirmed that fact.

You can see the sudden increase in Family numbers through the 10 million years after the K–T mass-extinction event, and see it reaching a maximum of 172 Families during the Miocene. Although they were very slow to diversify, prevented by the ecosystem and the Mesozoic dinosaurs, the curve seemed to be following the normal course of change. But the last four records, from 10, 5, 2 and 0.01 million years ago, show a big drop in Family numbers, which must mean that mammal Family diversity has passed its peak.

We tested these results against our spindle-curve model. We saw that mammal diversity peaked during the Miocene and that it is falling at the predicted slow rate. The solid line in figure 5.5 shows these calculations from the model, and it follows the spindle-shape theory precisely. If the curve is allowed to continue without interruption, without interference from outside the system, mammals will finally become extinct in about 900 million years' time.

This estimate is discerned from the slow fall in diversity once the Miocene peak is passed. You see the same long-drawn-out end to the bell curve in other groups, in the angiosperms for instance (see figure 5.4). Our projection for dinosaur Family diversity shows another theoretical long decline to extinction, which we know was stopped by the K–T event (see figure 2.3). But something is not quite right with the mammal curve in figure 5.5 and has not been seen before in these diversification curves. Those last four records at 10, 5, 2 and 0.01 million years ago show a very fast decline from 160 Families to 124 in just 10 million years. It's easy to explain why the number of mammal Families peaked during the Miocene, because temperatures and ecological harmony were also at their maxima then. Food was plentiful and there were no serious predators. But the sudden fall in numbers 10 million years ago is another matter.

After the Miocene peak the climate began to cool, leading into the ice ages of the Pleistocene. Then comes the puzzle! Why should the last four Family numbers have fallen so unusually quickly? As you can see in figure 5.5 that influence on the model curve is severe. A substantial proportion of Families became extinct in just a few thousand years. Does this mean that interference from outside the system had already begun?

With just four records the argument was ill-founded and needed confirmation from as many other data sources as possible. So we went on a search for more data, in the pessimistic knowledge that paleontologists are not renowned for preparing large sets of data, let alone sharing them. From Africa we found nothing other than different records of primates from some well-worked localities in Kenya and Tanzania. In Asia and Europe there are lots of species lists of mammals from particular regions or periods of geological time. All these data have confused and incompatible names, some recorded as species, others only as larger groupings. In Europe, Africa and Asia different techniques give different ages for the same specimens; controversy reigns. The specialists we consulted didn't have any easy solutions. We were stuck.

In the search for the right kind of database I knew we had to be very careful indeed, entering a field about which we know very little. So I read and consulted specialists, and soon found what I was looking for. It is a book published in 1980 by Björn Kurtén and Elaine Anderson, with an appendix listing all the known extinct and living species of mammals from the Pleistocene ice ages of North America. What's more, their presence or absence is recorded at 12 intervals of time over the last 2 million years. Excitedly, we set about transcribing the data from the appendix into an Excel spreadsheet. Two interglacials ago there were 34 Families of mammals in North America. Now there are ten. There were huge losses of mammal species over the same time, 335 species became extinct and only 210 survive. Over the 300,000 years involved that's some catastrophe, regardless of the fact that we are not too sure what the Family and species categories really mean.

In North America, the quality and quantity of knowledge about fossil mammals is very different from that in other parts of the world. There is more or less a single culture among the specialists, so the problems of naming and describing are much less severe than in Europe. There is also tremendous interest there in early man, not least for how human evolution and migration impinged on the new immigrant whites from Europe and blacks from Africa. Another reason for the rich knowledge is that there were an awful lot of mammals, species and individuals, in North America until very recently, famously populating the great plains leading up to the ice ages. We downloaded these data, compiled by John Alroy, from the Internet.

Our next challenge was to check the figures and set them beside other data to get some idea of their reliability. The changes in Family numbers at every one-million-year interval from these two databases are shown in figure 7.1. There were 53 known Families in North America at the end of the Eocene; in the whole world the peak was reached much later – 172 known in the Miocene. The four records through the last 300,000 years, from the appendix to the North American Pleistocene database, have 33, 34, 25 and 10 Families. It was a rapid and far-reaching fall.

We had found quantitative evidence that mammals are becoming extinct very quickly. What's more, the size and speed of the event is of mass-extinction proportions. The comparison with our earlier curve in figure 5.5 from the *Fossil Record 2* database is startling because those data did not include Pleistocene or present-day records.

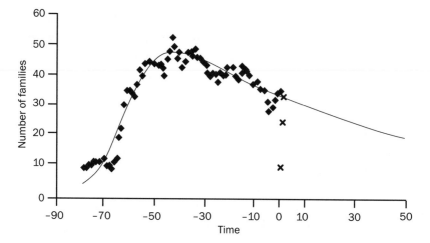

Fig. 7.1 The number of Families of North American mammals occurring every one million years from the Tertiary (diamonds are after John Alroy's data) and the Pleistocene (crosses are after Kurten & Anderson's data).
The solid line predicts the future extinction pattern without external influences after the Tertiary. The three crosses deflect from this trajectory.
(original compilation)

The few data from North American fossil sites only a few thousand years old change the interpretation from figure 5.5 appreciably. That curve used our bell-shape model to predict mammal extinctions as far away, 900 million years from now. Does it appear that something from outside nature has influenced the self-organised system of mammal evolution? Something has caused an avalanche.

The implications of the sudden large drop in records from these three sources of data are very important. We have only seen sudden falls like these during mass-extinction events. A simple comparison of the changes in the mammals with what happened with the dinosaurs (figure 2.3) shows two similarities. Both dinosaurs and mammals had reached peak Family numbers a few million years before the extinctions began. During the next few million years the curves follow our model with a fall in Family numbers. For both dinosaurs and mammals these early extinctions are of Families which were part of the first radiations of both groups and had become incompatible with the later forms. The second similarity is that after this steady fall from maximum diversity, both dinosaur and mammal extinctions go on to show up on the timescales of our curves as sudden vertical falls. The only times in the fossil record that show large and sudden falls like these are at mass-extinction events.

Of the two most likely causes of these sudden extinctions, climate change and the influence of humans, the first can have had only marginal influence. The climatic oscillations had been taking place for millions of years and showed no special change 12,000 years ago. Nothing special happened then, just the same cycles of warming and cooling and all the associated factors we know for the Pleistocene ice ages. The second likely cause, the influence of man, has a lot more support from what the paleoanthropologists are finding in North America and elsewhere. Then there are the effects of the rise of industrial processes to take into account.

The really frightening thing about this conclusion that we are living through a mass-extinction event, is that it is based on analysis of fossil data from specimens about 300,000 to 8,000 years old. The fossils provide statistically significant evidence that about a third of the Families of mammals have become extinct. What's more, modern humans' aggression has been the root cause of this present mass-extinction event that began at the end of the last ice age. It may have first been manifest with our killing off most of the large mammals from Africa, Asia and Europe, then perhaps the Neanderthals, and most recently game from the Americas. In Europe and Asia there are fewer clues to the killings, though similar cullings were going on.

Now there is the next wave of abuse from modern man that I discussed in the previous chapter, ultimately caused by our selfish burning of fossil fuels. So far, this process of human aggression hasn't begun to influence the curves for human extinction. Although the scene is set to make it inevitable, the catastrophe has not yet taken place.

And the duration is so short that the effect from events of the last 200 years since the industrial revolution are hard to put on the geological timescale. Nevertheless there is no shortage of speculation of how or when it might happen: just open a newspaper and read for yourself.

A double whammy

Of course, we can't make precise predictions about the scope of this mass-extinction event that we are living in. So far it has shown up most obviously with the large mammals. Other animals and some plants are following fast, as I explained in the previous chapter. Crises through which we can't survive are soon to hit particular human groups. Too much or too little water, famine, disease, and war are just five kinds of crisis awaiting us sooner rather than later.

What evidence is there from previous mass-extinction events to hint at what might happen to us? One feature that shows up in our Family diversification curves across earlier extinction events is the 'double whammy', two peaks when you expected just one. But maybe it depends on how you define the groups.

The mass-extinction event at the Permian–Triassic boundary, 245 million years ago, caused the extinction of more species than ever before or since. Although all the existing phyla survived the event, most Families did not. For example, more than two thirds of amphibian Families became extinct as did those of the dominant vascular plants, the gymnosperms and ferns (see figure 5.6). In each of these three groups the first phase of extinctions was followed by a period of diversification. New Families of amphibians, gymnosperms and ferns evolved into the Mesozoic and Tertiary and were much more effective in the very different ecosystems. Nevertheless, in each of these groups one or two species of a Family survived throughout the Mesozoic and Tertiary to today. Without mass extinctions the tail of the curve to final extinction is certainly very long.

Our results (figure 5.6) mean that these three groups experienced two phases of diversification, one in the Paleozoic and the other later in the Mesozoic. The full importance of these two phases in diversification remains to be thought through, but they do seem to be common phenomena. It causes me to ask the inevitable question: with their resilience to human actions, could the majority of small mammals survive the present mass-extinction event? Just like the modern Families of the three ancient groups, the small mammals can adapt to a world without their larger relatives. Ecologically they are economical, physiologically they are well protected, and they show all signs of maintaining their diversity through the changes we are forcing. Their small size and modest territorial and dietary requirements lead to their success. It will be interesting to see the genetic similarities and differences between the first and second phases of each group when that work is done.

The extinction event that we are experiencing now could be leading to a second diversification starting when the large mammals, especially we humans, are gone. Already a good proportion of the Families of large mammals has become, or is becoming, extinct. So far, the facts follow the pattern of the three Permian–Triassic groups that recovered from their first phase of extinction. We need to model two separate spindle-shaped curves for these groups that fall roughly into large and small mammal groups.

New life for the Earth

What can science predict about the future of life on the Earth? Ecological theory and computer modelling do have things to say about how the world might be, and we have trends set by previous catastrophes that enable us to make several clear deductions.

In May 2001, sixty-three Academies of Science from all parts of the world issued a statement on 'global trends in climate change'. It confirms the 90 per cent certainty that average global surface temperatures will rise by between 1° and 6°C over the next 100 years. The source of the predictions is the Intergovernmental Panel on Climate Change. That group also predicts rising sea levels, more rain and drought in different countries, and adverse effects on agriculture, health and water resources, in the same 100-year timespan. Perhaps the previous section is not a fantasy after all.

There are no reasons to believe that the plants will be seriously affected by the catastrophe. We know from our computer model (figure 5.4) that the most prominent plants, the angiosperms, are only just peaking in their Family diversity. Many herbs have benefited from our agricultural practices, with open fields and gardens. They have given new species and varieties both naturally and artificially. With their habitats gone these cannot survive, but the conifers and ferns do well in resisting environmental stress and will have an early boost in diversification. As we have seen already, some species of the rare tropical conifers may not survive.

The most vulnerable vegetation is the warm-temperate shrub surrounding the Mediterranean, western California and all tropical and subtropical forest. If any survives the forthcoming catastrophe it is likely that its range of species will be significantly reduced. That, in turn, will change the ecological balances in these distinctive and complex ecosystems. We have no way of knowing what kinds of forests they might become.

Many of the mammals are particularly vulnerable to the kinds of changes humans are inflicting on their lifestyle, and it is much more complex to predict their fate. Already I have nominated the 'large' mammals as the most likely to continue to become extinct. That's because there is evidence from our modelling of the fossil record data and arguments from ecologists that larger body weights need more food and space than their neighbours. The even larger dinosaurs are a precedent for this same argument.

There is no good evidence with which to distinguish 'large' from 'small'. Of the youngest order of mammals, the Primates, who is to say whether any of the monkeys will survive? I guess that their size, lifespan and behaviour are in their favour, but that the loss of many of their environments will threaten many of the species. The rodents and carnivores will always find food as long as they are not alone.

So we are left with our computerised models from the *Fossil Record 2* database, projecting times of peak diversity in large plant and animal groups. Two trends are clear. The plants will stay and their habitats will become restored to a new order. The curves for most of the other animal groups show that they are past their peak, on the long descent to extinction millions of years from now. But there are two notable exceptions, both in a previously unpopular domain, the air.

The insects and the birds are still at the early stage of high diversification. Both radiations are much longer and more diverse than most other animal groups have been, and neither shows signs of reaching its maximum. Here are the signs of where the new life on the Earth will develop when the large mammals are gone. The smaller ones – the rodents, insectivores, carnivores and bats – will be delighted. The far-stretched terrestrial ecosystems will be restored to new equilibria and the air will offer new space for innovation and developing communities.

It's a hollow feeling to think of this Earth without humanity: vast tracts of beauty without our eyes to behold. More difficult for the biologist to contemplate is the fact that the most complex group of creatures ever to have evolved has fallen into extinction. We have grown up to give reverence to the advanced characters of *Homo sapiens*, and now we know they were flawed. It seems that the largest genomes, the most complex physiology and neurology don't guarantee a permanent place on the throne of biodiversity. What we naïvely saw as an evolving hierarchy does not have ourselves, the human race, in its upper branches. The whole tree needs equal respect for all its parts.

Notes

CHAPTER I

Our manuscript about modelling evolutionary change was published as:

Hewzulla, D., Boulter, M. C., Benton, M. J. & Halley, J. M., 'Evolutionary patterns from mass originations and mass extinctions', *Philosophical Transactions of the Royal Society B* (1999), pp. 354, 463–9.

Northern temperate landscapes over the last 10 million years are described and discussed in:

Flannery, T., *The Eternal frontier* (New York: Atlantic Monthly Press, 2001). Boulter, M. C. & Fisher, H. C. (eds.), *Cenozoic Plants and Climates of the Arctic* (Berlin: Springer Verlag, 1994).

Vegetation, archaeology and anthropology of the developing Scottish landscape are reviewed in:

Birks, H. J. B., *Past and Present Vegetation of the Isle of Skye* (Cambridge: Cambridge University Press, 1973). Macnab, P. A., *The Isle of Mull* (Newton Abbot: David and Charles, 1970).

Major progress in biology through the first half of the last century is summarised in: Huxley, J. (ed.), *The New Systematics* (Oxford: Oxford University Press, 1940).

Concepts of geological time and physical processes on the Earth and in space are found in:

Harland, W. B. *et al.*, *A Geological Time Scale* (Cambridge: Cambridge University Press, 1990).
Lamb, S. & Sington, D., *Earth Story* (London: BBC Books, 1998).

Modern reviews of the possibilities of sustainable global management were written by European experts for a conference in Tokyo. See Heap, B. & Kent, J. (eds.), *Towards Sustainable Consumption* (London: Royal Society, 2000).

Seven comprehensive review articles on biodiversity make up an appendix, 'Nature Insight', in *Nature* 405 (May 2000), 208–54.

Other recommended reading:

Briggs, D. E. G., Erwin, D. H. & Collier, F. J., *The Fossils of the Burgess Shale* (Washington DC: Smithsonian Institute, 1994).
Desmond, A. & Moore, J., *Darwin* (London: Michael Joseph, 1991).
Kauffman, S., *At Home in the Universe* (Harmondsworth: Penguin, 1995). The big bang was 15 billion years ago and the universe is still expanding.
Levin, S. *Fragile Dominion: Complexity and the Commons* (Reading MA: Perseus, 1999).
Margulis, L. & Sagan, D., *Slanted Truths* (New York: Copernicus, 1997).
Rudwick, M. J. S., *The Great Devonian Controversy* (Chicago: Chicago University Press, 1965).
Ward, P., *The End of Evolution* (London: Phoenix, 1995).

CHAPTER 2

Jurassic paleoecology and landscapes are described and discussed by:

Frakes, A., Francis, J. E. & Syktus, J. L. *Climate Modes of the Phanerozoic* (Cambridge: Cambridge University Press, 1992).
Hallam, A. & Wignall, P. B., *Mass Extinctions and their Aftermath* (Oxford: Oxford University Press, 1997).

The Dorset coast:

Lambert, D., *The Ultimate Dinosaur Book* (London: Dorling Kindersley, 1993).

Late Cretaceous and Early Tertiary climate reconstructions from pollen and spore data are provided by Boulter, M. C., Gee, D. & Fisher, H. C., 'Angiosperm radiations at the Cenomanian/Turonian and Cretaceous/Tertiary boundaries', *Cretaceous Research* 19 (1998), pp. 107–12.

Cretaceous–Tertiary catastrophe:

Courtillot, V., *Evolutionary Catastrophes: the Science of Mass Extinction* (Cambridge: Cambridge University Press, 1999).
Eldredge, N., *The Pattern of Evolution* (New York: Freeman, 1998).

The P–Tr boundary event:

Erwin, D., *The Permian–Triassic Boundary* (New York: Scientific American, 1992).
Kerr, R. A., 'Whiff of gas points to impact mass extinction', *Science* 291 (2001), pp. 1469–70.

Further reading:

Ridley, M., *Evolution* (Oxford: Blackwell, 1998).
Sanz, J. L., 'An Early Cretaceous pellet', *Nature* 409 (2001), pp. 998–9.
www.ipcc.ch/pub/spm 19-02.pdf
Sepkoski, J. 'Ten years in the library: new data confirm paleontological patterns', *Paleobiology* 19 (1993), pp. 43–51.

CHAPTER 3

The Big Bang and early history of Earth:

Lamb, S. & Sington, D., *Earth Story* (London: BBC Books, 1998).
Hawking, S. W., *A Brief History of Time* (London: Bantam, 1988).

Cohen, B. A., Swindle, T. D. & Kring, D. A., 'Support for the lunar cataclysm hypothesis from lunar meteorite impact melt ages', *Science* 290 (2000), pp. 1754–6.

Sand piles and self-organised systems:

Bak, P., *How Nature Works: the Science of Self-Organized Criticality* (New York: Copernicus, 1996).
Buchanan, M., *Ubiquity* (London: Weidenfeld & Nicolson, 2000).
Halley, J. M., 'Ecology, evolution and l/f noise', *Trends in Evolution and Ecology* 11 (1996), pp. 33–7.
Johnson, S., *Emergence: The Connected lives of Ants, Brains, Cities and Software* (London: Allen Lane, 2001).
A collection of papers entitled Complex Systems. *Nature Insight*, 8 March 2001, 410, pp. 241–84.

Punctuated equilibria and logistic models:

Courtillot, V., *Evolutionary Catastrophes: The Science of Mass Extinction* (Cambridge: Cambridge University Press, 1999).
Eldredge, N., *The Pattern of Evolution* (New York: Freeman, 1999).
Benton, M. J., 'Models for the diversification of life', *Trends in Evolution and Ecology* 12 (1997), 490–5.

Mass extinctions:

Frakes, L. A., Francis, J. E. & Syktus, J. L., *Climate Modes of the Phanerozoic* (Cambridge: Cambridge University Press, 1992).
Hallam, A. & Wignall, P. B., *Mass Extinctions and Their Aftermath* (Oxford: Oxford University Press, 1997).
Hewzulla, D., Boulter, M. C., Benton, M. J. & Halley, J. M., 'Evolutionary patterns from mass originations and mass extinctions', *Philosophical Transactions of the Royal Society London B* 354 (1999), pp. 463–9.
Kirchner, J. W. & Weil, A., 'Delayed biological recovery from extinction throughout the fossil record', *Nature* 404 (2000), pp. 177–80.

Gaia:

Margulis, L. and Sagan, D., *Slanted Truths* (New York: Copernicus, 1997).
Lenton, T. M., 'Gaia and natural selection', *Nature* 394 (1998), pp. 439–47.
Lovelock, J., *Homage to Gaia, the Life of an Independent Scientist* (Oxford: Oxford University Press: 2000).

CHAPTER 4

K–T and the recovery:

Bains, S., Corfield, R. M. & Norris, R. D., 'Mechanisms of climate warming at the end of the Paleocene', *Science* 285 (1999), pp. 724–6.
Mukhopadhyay, S., Farley, K. A. & Montanari, A., 'A short duration of the Cretaceous–Tertiary boundary event: evidence from extraterrestrial helium-3', *Science* 291 (2001), pp. 1952–5.
Pfefferkorn, H. W., 'Recuperation from mass extinctions', *Proc. Nat. Acad. Sci.* 96 (1999), pp. 13597–9.

Opening of the North Atlantic:

Boulter, M. C. & Kvacek, Z., *The Palaeocene Flora of the Isle of Mull*, Special Paper in Palaeontology 42 (London: Palaeontological Association, 1989).
Boulter, M. C. & Manum, S., 'A lost continent in a temperate Arctic', *Endeavour* 21 (1997), pp. 105–8.

Tertiary cooling:

Boulter, M. C. & Fisher, H. C. (ed.), *Cenozoic Plants and Climates of the Arctic* (Berlin: Springer Verlag, 1994).
Flannery, T., *The Eternal Frontier. An Ecological History of North America and Its Peoples* (New York: Atlantic Monthly Press, 2001).
Kerr, R. A., 'How grasses got the upper hand', *Science* 293 (2001), pp. 1572–3.

Icehouse world:

Bennett, K. D., 1997. *Evolution and Ecology* (Cambridge: Cambridge University Press, 1997).
Betancourt, J. L., 'The Amazon reveals its secrets – partly', *Science* 290 (2000), pp. 2274–5.
Blunier, T. & Brook, E. J., 'Timing of millennial-scale climate change in Antarctica and Greenland during the last glacial period', *Science* 291 (2001), pp. 109–10.
Fagan, B., *The Little Ice Age* (New York: Basic Books, 2000).
Hillaire-Mercel, C. *et al.*, 'Absence of deep-water formation in the Labrador Sea during the last interglacial period', *Nature* 410 (2001), pp. 1073–7.
Schrag, D. P., 'Of ice and elephants', *Nature* 404 (2000), pp. 23–4.

CHAPTER 5

Species and other taxa:

Benton, M. J., *Vertebrate Palaeontology* (London: Chapman and Hall, 1997).
Koerner, L., *Linnaeus – Nature and Nation* (Cambridge, MA: Harvard University Press, 1999).

Ideas on evolution since Darwin:

Gee, H., *Deep Time* (London: Fourth Estate, 2000).
Gould, S. J., *Eight Little Piggies* (Harmondsworth: Penguin, 1994), essay about Louis Agassiz, 'A Tale of Three Pictures', pp. 427–38.
Stafleu, F. A., *Linnaeus and the Linnaeans* (Utrecht: A. Oosthoek's Uitgeversmaatschoppij, 1971).
Stewart, W. N., *Paleobotany and the Evolution of Plants* (Cambridge: Cambridge University Press, 1983).
Wieser, W., 'A major transition in Darwinism', *Trends in Ecology and Evolution* 12 (1997), pp. 367–70.

Bell curves:

Poole, R., *Towards Deep Subjectivity* (London: Allen Lane, 1972).
Boulter, M. C. and Hewzulla, D. 1999. For every extinct group singled out
from *The Fossil Record 2* and our other databases, our plots give curves that
follow our model. You can try yourself on an interactive website (*http://
www.erdw.ethz.ch/_pe/1999_2/model/issue2_99.htm*) where our scientific
article about this work in the new electronic journal *Palaeontologica Electronica*
is published. The consistent results merely confirm what so many evolutionary
biologists have agreed for so long, that the shape of Agassiz's spindle is an
accurate portrayal of the origin, diversification and extinction of large clades
of animals and plants.

CHAPTER 6

Neanderthal man:

Ponce de Leon, M. S. & Zollikofer, C. P. E., 'Neanderthal cranial ontogeny
and its implications for late hominid diversity', *Nature* 412 (2001), pp. 534–7.
Stringer, C. & Davies, W., 'Those elusive Neanderthals', *Nature* 413 (2001),
pp. 791–2.

Conservation:

Conservation is a response to protect habitats from progressive destruction.
It attempts to manage an ecosystem, even in an artificial way, so as to maintain
as many features of the working parts as possible.
Houghton, J. T. *et al.*, *Climate Change 2001: The Scientific Basis. Third IPCC
Assessment* (Cambridge: Cambridge University Press, 2001).
Heap, B. & Kent, J., *Towards Sustainable Consumption: a European Perspective*
(London: Royal Society, 2000).
Smith, J. & Uppenbrink, J. (ed.), 'Earth's variable climatic past', *Science* 292
(2001), pp. 657–93 – a special report with five reviews.
Whittaker, R. J., *Island Biogeography* (Oxford: Oxford University Press, 1998).

Climate change:

Calder, N., *The Manic Sun – Weather Theories Confounded* (London: Pilkington Press, 1997). This view is based on data from the 1995 IPCC reports on clouds, sunspots, atmosphere and ozone levels.

Hillaire-Marcel, C. *et al.*, 'Absence of deep-water formation in the Labrador Sea during the last interglacial period', *Nature* 410 (2001), pp. 1073–7.

Lomborg, B., *The Sceptical Environmentalist* (Cambridge: Cambridge University Press, 2001). Green beliefs are discarded in favour of the author's argument that the planet is doing better than ever.

CHAPTER 7

Evolutionary psychology:

Blackmore, S., *The Meme Machine* (Oxford: Oxford University Press, 1999).
Pinker, S., *How the Mind Works* (London: Allen Lane, 1997).
Rose, H. & Rose, S. (eds.), *Alas, Poor Darwin* (London: Cape, 2000).
Wilson, E. O., *Consilience* (London: Little, Brown, 1995).

Mammals and early humans:

The two databases of mammal occurrences in North America can be found at *biodiversity.org.uk* by clicking on 'search data' and following the instructions.

Alroy, J., 'A multispecies overkill simulation of the end-pleistocene mega-faunal mass-extinction', *Science* 292 (2001), pp. 1893–6.

Boulter, M. C. & Hewzulla, D., 'Evolutionary modelling from Family diversity', *Palaeontologia Electronica* 2, 2 (1999), *www-odp.tamu.edu/paleo*.

Flannery, T., *The Eternal Frontier* (London: Heinemann, 2001).

Kurten, B. & Anderson, E., *Pleistocene Mammals of North America* (New York: Columbia University Press, 1980).

Tattersall, I., 'Once we were not alone', *Scientific American* (January 2000), pp. 38–44.

Patterns and beauty:

Jencks, C., 'The four principles of beauty', *Prospect* (August 2001), pp. 22–7.
Midgley, M., *Science and Poetry* (London: Routledge, 2001).
Poole, R., *Towards Deep Subjectivity* (London: Allen Lane, 1972).
Stewart, I., *Life's Other Secret* (Harmondsworth: Penguin, 1998).

New life from extinctions:

Buchanan, M., *Ubiquity* (London: Weidenfeld & Nicolson, 2000).
Levin, S., *Fragile Dominion* (Reading, MA, 1999). Argues that self-organisation is more widespread than most ecologists have anticipated. When the variations within a system are small, modularly organised webs of interaction become the determining force. Already in this new century there are many new examples of systems with self-organised power laws and pink noise.

Acknowledgements

Like so many concoctions, this project originated around a dining table. There were many Friday evenings of argument with the novelist Vincent Brome, the late Dr Charles Rycroft, the psychoanalyst, the inimitable Colin Merton and myself. It was to start as a magazine article but has ended up as a book.

I also thank Vincent for encouraging me through the many ups and downs that a first-time author suffers to find a publisher. There were many others who advised me at the beginning on literary matters including Alan Cameron, Hilary Rubenstein, Nigel Calder and Jason Cooper.

At the University of East London, my research group with Dr Dilshat Hewzulla, Richard Hubbard, David Gee and Helen Fisher has been a wonderful stimulus and Professors Keith Snow and David Edwards have given invaluable support. Dr David Polly at Queen Mary's College down the road helped with the mammals, as did my undergraduate student Filippo Campagna Popol. Dr Roger Poole at the University of Nottingham gave useful comments on some of the philosophical questions. Dr John Alroy at Santa Barbara, Professor Mike Benton at Bristol and Dr Ken Piel in Massachusetts compiled some of the databases. Ian McKay, University of Glasgow, advised me on Tirefour broch.

My doctors and nurses at the Royal Free Hospital helped make the last stages of writing comfortable and productive.

Professor Svein Manum, Institute of Geology, University of Oslo and Dr Raphie Kaplinski, University of Sussex, have read through the manuscript and offered many suggestions for improvement. Professors

Bill Chaloner and John Charap, University of London, made many comments to improve chapters 5 and 3 respectively. I am responsible for the errors and weaknesses that remain.

At Fourth Estate Clive Priddle showed me how to improve the first version. My editor Leo Hollis has been a wonderfully constructive guide to help the ideas unfold. Without him the project would not have happened. The copy-editor, Steve Cox, made many valuable suggestions for improvement.

Biddy Arnott has been an enthusiastic and patient wife to discuss and offer improvements, and our boys, Tom and Alex, have been intrigued. Importantly, in so many ways all these people and many more have helped make it be fun.

Index

Page numbers in *italic* refer to the illustrations

Agassiz, Louis 130, 132–4, *133*, 135, 147,
    148, 149
Agnatha 150–1, *150*
agriculture 169, 191
Ahat, Alim 16
algae 46
alien species, introduction of 168–9
Alroy, John 160
Alvarez, Walter and Luis 40–2, 89
ammonites 24, 25, 27, 28, 40, 43, 44, 89
amphibians *155*, 190–1
Anderson, Elaine 186–7
angiosperms (flowering plants) 45, 95
    diversification curve 151–2, *152*, 192
    evolution 36–9, 93
    pollen database 93–5
Antarctica 111, 120–1, *175*
Arctic Ocean 106
Aristotle 126–7, 128–9, 130, 136
asteroids, moon event 61
    see also meteorites
Atlantic Ocean
    climate change 122–3, 171–5
    formation of 98–100, 103, 106
    Gulf Stream 1, 99, 105–6, 110, 120, 121,
        122, 172
    lost continent in 110–12
    North Atlantic Conveyor 120, 172–4, *175*
    volcanoes 99, 100–1, 103, 107
Atlantis 109–11

atmosphere
    after K–T event 88
    COFD concentration 115–16
    greenhouse effect 116
    volcanic changes 54
avalanche effect, self-organised systems 63, 64,
    64, 82, 156, 182, 183

Bak, Per 63, 65, 67, 75, 82, 87, 124, 150
basalt 100, 101, *102*, 103
behaviourism 177–82
Benton, Mike 76, 80–1, 82, 85, 86
Big Bang 56–7, 60, 62
Big Five extinction events 50–2, *51*, 75, 80,
    82, 85, 86, 170
biodiversity 18–22
    hotspots 20
    interdependence 33–4
    Jurassic period 29–30
    monitoring 19–22
    stable state 34
    'survival of the fittest' 30, *31*, 72, 146
    see also diversification
birds 24–5
    after K–T event 91
    dispersal of tree seeds 167
    evolution 91–2, 154
    extinctions 168, 169
    survival of 193
Birks, John 6

body temperature xiii
brain, human 177
Briggs, Derek 11
Brito-Arctic Igneous Province 101, 104–5
buffalo 122–3
Buffon, Georges 129
the Burren 5

carbon dioxide, in atmosphere 13, 116, 117
Caribbean Sea 166
catastrophes
  Big Five extinction events 50–2, 51, 75, 80, 82, 85, 86, 170
  current mass-extinction event 187–93
  Cuvier's theory of catastrophism 133–4
  environmental change and 47
  essential to evolution 62, 82, 183
  K–T event 39–45, 48, 56, 67, 86, 88–93
  moon event 61–2, 68
  P–Tr event 54–5, 56, 85, 86, 89, 190
  26-million-year cycle theory 48–51, 72–3, 75, 80
Chalmers, Neil 143, 144–5
chaos theory 57–60, 184
China 16, 18
Cimolesta 151, 151
cladistics 139, 140, 145–6
classification 126–30, 134–5, 136, 139–40
clearances, eighteenth-century 7
climate change 162–5, 191–2
  after K–T event 94, 119
  atmospheric COFD and 115–16
  Eocene 108–9
  evidence from ice cores 120–2
  and extinctions 161–2
  human activity and 182–3
  human adaptation to 163–4, 174–6
  Jurassic period 31–3
  'Little Ice Age' 122–3, 163
  and mammal extinctions 189
  melting ice 174–5
  in Miocene 112–17
  North Atlantic Conveyor 120, 172–4, 175
  rainfall 5–6
  seasonal changes 62, 68

Tertiary period 97, 163
Younger Dryas Event 121–2, 123
  see also ice ages
coal 6, 10–11
conifers 38, 45, 95, 111, 114, 164, 192
continental drift 28, 29, 52–3, 53, 98, 99, 101–3, 106, 110
Courtillot, Vincent 81, 84, 87
creationism 132, 133–4, 136, 146
Cretaceous period 25–6, 148
Croll, James 118, 119
C–T boundary 36–7
Cuvier, Georges 130, 132, 133–4

'Daisyworld' 66
Dark Ages 121
Dartmoor 35
Darwin, Charles 30, 34, 35, 69–70, 80, 132
  evolutionary trees 70–1, 71, 72, 134–5, 139
  natural selection 130, 131, 133, 135, 137
  The Origin of Species 49, 70, 131, 134, 135, 137, 178
  preadaptation 47
databases
  DNA 180
  fossils 73–4, 75, 76–82, 84–7, 150
Dawkins, Richard 177–80, 181
dinosaurs
  and birds 91–2, 153
  decline in diversity 43–4, 44, 45, 185–6, 188–90
  extinction xii, 39, 40, 43–4
  as fairy stories 23, 26
  food chain 25
diversification
  after K–T event 90–1, 95–6
  decline in 43–4, 44, 185–9
environmental change and 147–8
  K–T catastrophe 47
  punctuated equilibrium theory 81
  'spindle diagrams' 132–3, 133, 134, 147, 148, 149–56
  see also biodiversity
DNA 72, 178

databases xi, 139, 145, 180
evolution of humans 158
K–T catastrophe 46–7
relationship with ecology 131
dodo 168

ecology 131, 136–7
see also environmental change
ecosystems, as self-organised system 124
Edwards, W.N. 104, 105
Egypt 1–3
Eldredge, Niles 74–5, 79, 80, 81, 89
elephant bird 168
elk, Irish 167–8
Enlightenment 127, 129
environmental change
    as catalyst for evolution 147
    human activity and 9–11, 165–7,
        182–3
    and human extinction 182
    and mass-extinction events 47
    26-million-year cycle theory 48–51, 72–3,
        75, 80
Eocene epoch 13, 106, 108–9, 114
evolution 12–14
    and adaptability 26, 27
    after K–T event 89–98
    catastrophes essential to 62, 82, 183
    as cellular process 130–1
    and classification 134–5
    environmental change and 147
    evolutionary psychology 177–82
    evolutionary trees 70–1, 71, 72, 134–5,
        138, 139–40, 146
    exponential change 79–87, 83, 91–2, 92,
        153, 156
    flowering plants 36–9, 45
    genetics 131, 136–8
    Jurassic period 25–7
    K–T catastrophe 45–7
    mechanism of 146
    migration and 162
    naming fossil species 140–3
    natural selection 130–1, 133, 137, 146,
        178

punctuated equilibrium theory 74–5, 79,
    81
'spindle diagrams' 132–3, 133, 134, 147,
    148, 149–56
visualising evolutionary change 143–8
extinction events *see* catastrophes

fairy stories, dinosaurs as 23, 26
ferns 93–4, 97, 100, 155, 190–1, 192
fire, K–T catastrophe 39–40
fish, overfishing 108, 166
fitness landscapes 30–1, 31, 33
Flood, 132, 134
floods 174, 175
flowering plants *see* angiosperms
food chain, Jurassic period 25
forests
    deforestation 7, 11, 38, 166–7
    fossils ix–xi
    in Miocene 115
    Tertiary period 96–8
fossils 11–12
    databases 73–4, 75, 76–82, 84–7, 150
    defining extinct species 138
    early humans 157–9
    interglacial periods 8
    Lazarus taxa 50
    naming species 140–3
    peat 6–8
fractals 59, 68
fuels, fossil 10–11, 189

Gaia theory 65–7
Galapagos Islands 69–70, 169
Gardner, Starkie 100, 104
gas, natural 95
Gaudemar, Yves 81
Gee, David 19
Gee, Henry 145
genetics
    DNA databases xi, 140, 145, 180
    evolution 130–1, 136–8
    and human behaviour 177–82
    'survival of the fittest' 30, 31
Genyornis 168

geology
  environmental change and 27–8, 34–8
  K–T catastrophe 40–3
  moon 61
  timescale 11–15
Giant's Causeway 101, *102*
glaciers x, 121–2, 174
global warming 37–8, 55, 172, 174, 182–3,
  191–2
GM crops 169
Godwin, Professor Sir Harry 117, 118
Gould, Stephen Jay 11, 47, 74, 75, 79, 80,
  81, 87, 89, 135
grasses 114–16, 122
greenhouse effect 13, 53, 54, 55, 96–7, 115,
  116
Greenland 99, 120–1, 122, 174, 175
Gulf Stream 1, 99, 105–6, 110, 120, 121,
  122, 172
gymnosperms 190–1

Haldane, J.B.S. 30, 136, 156
Hamilton, William 66
Hawking, Stephen 59
Hewzulla, Dilshat 16, 21, 86–7, 138, 148–9
Himalayas 174
Homo erectus 15
Homo habilis 15
Hovgaard Ridge 111
Hoyle, Fred 4
Hubbard, Richard 16, 32
Hughes, Norman 141, 142
humans
  aggression 8–9, 157–62, 164, 176, 182,
    189
  environmental change 9–11, 165–7, 182–3
  evolution of 15
  evolutionary psychology 177–82
  extinction of 170, 182, 185, 190, 193
  and extinction of large mammals 8–9,
    160–1, 162, 189
  fossil record 157–9
  introduction of alien species by 168–9
  Iron Age 1–5
  language 159

migration 157, 158, 162, 163–4, 174
population growth 165
selfishness 180, 182, 189

ice
  climate change 174–5
  ice cores 120–1
ice ages xiv, 15, 101, 189
  causes 117–20
  in future 171–4
  large mammal extinctions 8–9
India xiii, 106
Industrial Revolution 10, 163, 165, 180, 182
insects 193
iridium 40, 41, 42, 89
Iron Age 1–5
island extinctions 161, 165, 167–8, 169

JOIDES Resolution 103, 110
Jurassic period 23–30, *29*, 31–3
Jussieu, Antoine 130

K–T extinction event 39–45, 48, 56, 67, 86,
  88–93
Kauffman, Stuart 63, 65, 67
Kirchner, James 66–7, 87
Komarov Botanical Institute 78
Kukla, George 171, 172
Kurtén, Björn 186–7
Kvacek, Zlatko 105

Labrador Sea 172, 174
Lamarck, Jean 129–30
language 159
life
  future of life on Earth 191–3
  as self-organised system 67
  see also evolution
Linnaeus, Carl 127–8, 129, 130, 134, 135,
  136, 139–40
Linnean Society 135, 139–40
'Little Ice Age' 122–3, 163
Lovelock, James 65–6, 67
Lyell, Charles 21, 38, 51, 70, 135, 141–2
Lyme Regis 23–4, 28, 29, 31–2, 55

mammals
browsers and grazers 114–16
decline in diversity 185–9, *188*
diversification 92, 95–6, 97, 151–2, *153*
extinction of large mammals 8–9, 160–1, 162, 182
future of 185, 188, 191, 192
island extinctions 167–8
K–T catastrophe 46, 90
'Man and the Biosphere' (MAB) 19–20
Manum, Svein 103, 109–10
Martin, Paul 160
mass-extinction events *see* catastrophes
May, Sir Robert xiii, 170
Mendel, Gregor 131
Mesozoic period 35, 190, 191
meteorites xiii, 10, 12
K–T catastrophe 39–45, 88–9
P–Tr event 55, 56
26-million-year cycle theory 48–50
migration 157, 158, 161–2, 163–4, 174
Milankovitch, Milutin 118–19
Miocene epoch 113–16, 148, 185, 186
molecular biology 144
moon 61–2, 68, 87
Mull, Isle of 98–104, 105

NASA 48, 87
Nathorst, A.G. 105
Natural History Museum, London 144–5
natural selection 130–1, 133, 137, 146, 178
Neanderthal man 9, 15, 157, 158–9, 164, 165, 189
Nee, Sean 170
Nemejc, Professor Frantisek 141, 143
El Niño 120, 121, 122, 172
North Atlantic Conveyor 120, 172–4, 175
nuclear weapons 43, 176
nuclear winter 48

oceans
after K–T event 90–1, 105–6
climate change and 171–4, 175
currents 52, 106, 120, 121, 122, 172–4
oxygen levels 37
plankton 46, 88–9, 90, 103–4, 107
rise in water temperature 106–7
sea level changes 55, 107–8, 109, 175, 191
oil 6, 10, 78, 85, 109
Oligocene epoch 98, 111–12, 148

P–Tr extinction event 54–5, 56, 85, 86, 89, 190
Pacific Ocean 61, 120, 121, 122, 171, 172
Paleocene epoch 12, 94, 97–8, 106
palms 113–14
Pangaea 28, *29*, 35
peat 6–8
Pelletier, Jon 124
photosynthesis 54, 96, 116
phytoplankton 88–9, 108
pink noise 59, 68, *69*, 86, 124
Pinker, Steven 177–80, 181
plankton 46, 88–9, 90, 103–4, 107
plants
extinctions 167
flowering plants 36–9, 45, 93–5, 96
future prospects 192, 193
K–T catastrophe 45
see also forests
plate tectonics 31, 52–4, *53*
Pleistocene epoch 15, 113, 118, 186, 189
pollen, fossil record 6–7, 93–4, 104–5, 110, 111, 112–13, 118
power laws 63, 64–5, *64*, 68, 75, 82, 86, 124, 183
psychology, evolutionary 177–82
pterosaurs 24
punctuated equilibrium theory, evolution 74–5, 79, 81

quartz 42–3

rainfall 5–6
rainforests 96, 97, 166–7
Rannoch Moor 7–8
Raup, David 48–9, 72–3, 74, 77, 79, 80
redwood trees 96, 113, 169–70
reef systems 166
rivers 108

sand piles, self-organised systems 63–5, *64*,
    68, 82, 84
Sanger, Fred xi
Santa Fe Institute 63, 65
Schopf, Jim 141
Scotland 1, 3–5, 6–8
sea level changes 55, 107–8, 109, 175, 191
self-organised systems 100, 149, 156
    avalanche effect 63, 64, *64*, 82, 156, 182,
        183
    catastrophes as necessary feature of 183
    ecosystems as 124
    Gaia theory 65–7
    'pink noise' 68
    sand piles 63–5, *64*, 68, 82, 84
Sepkoski, Jack 48–9, 50–1, *51*, 72–4, 75–6,
    77, 79, 80, 84, 85
Seward, Albert 104, 105
sexual reproduction 131
Shackleton, Nick 118
Siberia Man 9
Simpson, George Gaylord 30, 136
Smith, John Maynard 145
Smithsonian Institution 78
sparrows 169
species
    cladistics 139, 140, 145–6
    classification 128–30, 134–5, 136, 139–40
    definitions of 137, 139
    exponential change in populations 82, *83*
    naming fossils 140–3
    recently extinct species 165
    as self-organised system 123
    'spindle diagrams' 132–3, *133*, 134, 147,
        148, 149–56
    total number of 138–9, 148
    see also biodiversity; diversification
Species 2000 19
Spencer, Herbert 30
'spindle diagrams' 132–3, *133*, 134, 147,
    148, 149–56
Stewart, Ian 156

stromatolites 166
sun, Earth's orbit 119–20
'survival of the fittest' 30, *31*, 72, 146
Sylvester-Bradley, Peter 143

taxonomy *see* classification
teeth, grazers 115–16
temperatures
    icehouse world 116
    increase after K–T event 94–5, *119*
    oscillations in Eocene 108–9
    rise in ocean water temperature 106–7
Tertiary period 88–98, 163, 190
Tirefour Broch 1, *2*, 3–5
Trafalgar Square interglacial 8
trees *see* forests
Triassic period 25, 31–2, 55
trilobites 74–5

UK Systematics Association 49–50
UNESCO 19
uniformitarianism 21, 38, 142
United Nations 9–10, 18, 138
universe, emergence of structures 56–7, 59,
    60, 62

volcanoes 35
    Atlantic Ocean 99, 100–1, 103, 107
    C–T eruptions 37–8
    Deccan Traps xv, 106
    P–Tr event 54–5

Walcott, Charles 12
Wallace, Alfred Russel 135
water shortages 174
white noise 57–9, *58*, 67
Wieser, Wolfgang 146
Wilson, E.O. 178
Wright, Sewall 30, 136

Younger *Dryas* Event 121–2, 123
Yucatán Peninsula 39, 40, 42, 89

Reflections on Afro-American Music

Dominique-René de Lerma
Founder, The Black Music Center
Indiana University

with contributions from
Richard L. Abrams
Julian "Cannonball" Adderley
Samuel Akpabot
Thomas Jefferson Anderson
Dorothy Ashby
David N. Baker
John A. R. Blacking
Marian Brown
Earl Calloway
John Carter
Thomas A. Dorsey
Charles Ellison
Phyl Garland
Frank Gillis
Carole Johnson
Pearl Williams Jones
Robert H. Klotman
Johnnie V. Lee
Don Malin
Portia K. Maultsby
Undine S. Moore
John E. Price
Geneva Handy Southall
Herndon Spillman
John A. Taylor
Sharon B. Thompson
and others